Instructor's Resource Manual

for

An Introduction to Genetic Analysis

Sixth Edition

Anthony J. F. Griffiths
University of British Columbia

William M. Gelbart
Harvard University

W. H. Freeman and Company
New York

ISBN: 0-7167-3062-6

Printed in the United States of America.

Second printing 1997

Contents

Preface

The Instructor's Resource Manual has been prepared to accompany the textbook *An Introduction to Genetic Analysis*, sixth edition, and follows the chapter layout of that book. Each chapter contains a selection of both multiple-choice and open-ended problems. The open-ended problems and many of the multiple-choice questions involve data processing and inference, in keeping with the general analytical theme of the text. Furthermore, the nomenclature and symbols used in the problems follow the text's style. The types of problems and the general degree of difficulty are similar to the problems found at the ends of the text chapters, so the Test Bank should be useful for student assessment.

This is the first bank of problems prepared for *An Introduction to Genetic Analysis*, so the authors welcome input from users on problem design, solutions, and accuracy. Please send comments to W. H. Freeman by mail or e-mail:

W. H. Freeman and Company
Supplements Editor
41 Madison Avenue
New York, NY 10010

102263.2667@compuserve.com

We would like to thank Mary Forrest of Okanagan University College and Barbara Moon of the University College of the Fraser Valley for their generous help in preparing this manual.

Anthony Griffiths
William Gelbart

2

Mendelian Analysis

Multiple-Choice Questions

1.   A phenotypically normal woman is heterozygous for the recessive Mendelian allele causing phenylketonuria, a disease arising from the inability to process phenylalanine in food. What proportion of her eggs will carry the allele that allows normal processing of phenylalanine?
    a.   all
    b.   3/4
    c.   none
    d.   1/4
    e.   1/2 •

2.   A selfed monohybrid characteristically produces progeny phenotypes in the ratio
    a.   1:1.
    b.   1:2:1.
    c.   3:1. •
    d.   9:3:3:1.
    e.   1:1:1:1.

3.   A testcrossed monohybrid characteristically produces progeny phenotypes in the ratio
    a.   1:1. •
    b.   1:2:1.
    c.   3:1.
    d.   9:3:3:1.
    e.   1:1:1:1.

4.   If genes assort independently, a selfed dihybrid characteristically produces progeny phenotypes in the ratio
    a.   1:1.
    b.   1:2:1.
    c.   3:1.
    d.   9:3:3:1. •
    e.   1:1:1:1.

5.   If genes assort independently, a testcrossed dihybrid characteristically produces progeny phenotypes in the ratio
    a.   1:1.
    b.   1:2:1.
    c.   3:1.
    d.   9:3:3:1.
    e.   1:1:1:1. •

6.   In guinea pigs B = black and b = brown. In one lab the cross records of the parents had been lost but the progeny were all black. Certain pairs of progeny were mated repeatedly over a number of years, each cross producing numerous offspring. Some of these crosses produced only black progeny and some produced black and brown in a 3:1 ratio. The genotypes of the original parents must have been
    a. $BB \times Bb$. •
    b. $Bb \times Bb$.
    c. $BB \times BB$.
    d. $bb \times BB$.
    e. $bb \times Bb$.

7.   In hogs, a dominant allele B results in a white belt around the body. At a separate gene the dominant allele S causes fusion of the two parts of the normally cloven hoof resulting in a condition known as syndactyly. A belted syndactylous sow was crossed to an unbelted cloven-hoofed boar, and in the litter there were:
    25% belted syndactylous,
    25% belted cloven,
    25% unbelted syndactylous,
    25% unbelted cloven.
The genotypes of the parents can best be represented as
    a. $BB\ SS \times bb\ ss$.
    b. $Bb\ Ss \times bb\ SS$.
    c. $Bb\ Ss \times BB\ ss$.
    d. $bb\ Ss \times Bb\ ss$.
    e. $Bb\ Ss \times bb\ ss$. •

8.   If a plant of genotype *Aa Bb Cc Dd* is selfed and the genes assort independently, how many different genotypes will be found among the progeny?
    a.   81 •
    b.   16
    c.   4
    d.   24
    e.   64

9.   If two mice of genotype *Ff Gg Hh Ii Jj* are repeatedly mated, how many different phenotypes will be found in the progeny?
    a.   15
    b.   32 •
    c.   128
    d.   256
    e.   16

10.   In the cross female *Aa Bb cc Dd ee* × male *Aa bb Cc Dd*, what proportion of progeny will be phenotypically identical to the female parent? (Assume independent assortment of all genes.)
    a. 1/16
    b. 3/8
    c. 3/16
    d. 9/16
    e. 9/64 •

11.   The following pedigree concerns the autosomal recessive disease phenylketonuria (PKU). The couple marked A and B are contemplating having a baby but are concerned about the baby having PKU. What is the probability of the first child having PKU?

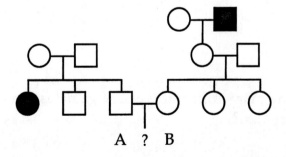

A   ?   B

    a.   0
    b.   1/4
    c.   1/12 •
    d.   3/4
    e.   9/64

12.   Mendel crossed *YY RR* (yellow wrinkled peas) with *yy rr* (green smooth peas) and selfed the F_1 to obtain an F_2. In the F_2 what proportion of yellow wrinkled individuals were pure-breeding?
    a.   9/16
    b.   1/16
    c.   1/4
    d.   3/16
    e.   1/9 •

13.   In corn, *P* = pigmented and *p* = unpigmented, and *T* = tall and *t* = dwarf. The two gene pairs are inherited independently. In a cross pigmented tall × pigmented dwarf, the progeny were:
    3/8 pigmented tall,
    3/8 pigmented dwarf,
    1/8 unpigmented tall,
    1/8 unpigmented dwarf.
The genotypes of the parents must have been
    a.   *Pp Tt × Pp Tt.*
    b.   *Pp TT × PP tt.*
    c.   *pp Tt × Pp tt.*
    d.   *Pp Tt × Pp tt.* •
    e.   *PP Tt × Pp tt.*

14.   A man and his wife are both heterozygous for the autosomal recessive allele for albinism. If they have two children, what is the probability that both children will be phenotypically identical with regard to skin color?
    a.   3/4
    b.   1/16
    c.   1/4
    d.   9/16
    e.   5/8 •

15.   In a trihybrid how many different types of gametes can be produced with regard to the three genes concerned?
    a.   3
    b.   4
    c.   8 •
    d.   27
    e.   16

16.   A couple are both heterozygous for <u>two</u> autosomal recessive diseases: cyctic fibrosis (CF) and phenylketonuria (PKU). What is the probability of their first child having either CF *or* PKU?

    a.   0
    b.   1/4
    c.   1/2 •
    d.   9/16
    e.   1/16

Open-Ended Questions

1.   A couple of normal ancestry have two normal children and an infant affected with Tay-Sachs disease (autosomal recessive). The sister of the husband wants to marry the brother of the wife; in such a mating, what would be the probability of their first child having Tay-Sachs disease?

Solution
The couple are heterozygous so in both sets of their phenotypically normal parents there must have been one heterozygote. Therefore both sets of grandparents must have been $Tt \times TT$ and the brother and sister each stand a 1/2 chance of being Tt, and if so then there is a 1/4 chance of an affected infant. Overall probability = $1/2 \times 1/2 \times 1/4 = 1/16$.

2.   A man whose uncle had cystic fibrosis (autosomal recessive disease) is married to a woman whose cousin had cystic fibrosis.

    a.   What is the probability that their first child will have cystic fibrosis?
    b.   If they have two children, what is the probability that they will both be phenotypically normal?

Solution
    a.   For the man's uncle to be tt the respective grandparents must have both been Tt, so there is a 2/3 chance that one of the man's parents is Tt, and a further 1/2 chance that the man would have received the t from that parent. The woman's parent's sibling must have been Tt so her grandparents were most likely $TT \times Tt$ and the woman's parent would then have a 1/2 chance of being Tt, and there would be a further 1/2 chance of passing the t on to the woman herself. Finally if both parents are Tt, there is a 1/4 of a tt child. Overall probability is $2/3 \times 1/2 \times 1/2 \times 1/2 \times 1/4 = 2/96 = 1/48$.
    b. $47/48 \times 47/48$

3. Consider the following pedigree of a rare autosomal recessive disease.

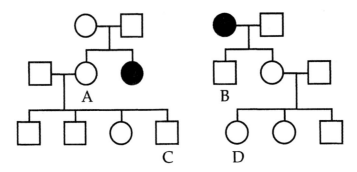

a. If individuals A and B marry, what is the probability that their first child will have the disease?

b. If individuals C and D marry, what is the probability that their first child will have the disease?

c. If the first child of marriage C × D is normal, what is the probability that their second child will have the disease?

d. If the first child of marriage C × D has the disease, what is the probability their second child will have the disease?

(Assume all people marrying into the pedigree do not carry the abnormal allele.)

Solution

a. Choosing *M* for unaffected and *m* for disorder, *b* must be *Mm*, and A has a 2/3 chance of being *Mm*. Overall chance of an affected child is $1 \times 2/3 \times 1/4 = 2/12 = 1/6$.

b. If C's mother 'A' is heterozygous, C stands a 1/2 chance of being heterozygous. D's mother must be heterozygous, and D stands a 1/2 chance of inheriting that hetrozygosity. Overall chance of an affected child is $2/3 \times 1/2 \times 1 \times 1/2 \times 1/4 = 2/48 = 1/24$

c. The probability is still 1/24.

d. Now we know individuals C and D must both be *Mm*, so the chance of the second child being *mm* is 1/4.

4. In peas, tall (*T*) is dominant to dwarf (*t*), and red flowers (*R*) is dominant to white flowers (*r*). A tall white-flowered plant was crossed with a dwarf red-flowered plant and 100 seeds were grown up into plants. There was found to be

    50 tall red-flowered plants and
    50 dwarf red-flowered plants.

a. List the genotypes of the parents and the two kinds of offspring.

b. If a dwarf red-flowered offspring plant was selfed, what progeny types are expected and in what proportions?

Solution

    a.   All progeny are red so parents must have been *rr* × *RR*. There is a 1:1 ratio of tall : dwarf so the parents must have been *Tt* × *tt*. Overall the parents were *Tt rr* and *tt RR*. The offspring were 1/2 *Tt Rr* and 1/2 *tt Rr*.

    b.   This individual must have been *tt Rr*, so selfing would give a ratio of 3/4 dwarf red and 1/4 dwarf white.

5.   The ability of the human body to break down the red color in beets is an autosomal dominant phenotype. The inability is recessive, detected by red coloration of the urine (we will call this phenotype 'secretor'). If a nonsecretor woman with a secretor father marries a nonsecretor man who in a previous marriage had a secretor daughter, what is the probability that their first child will be

    a.   a secretor girl,

    b.   a nonsecretor girl, or

    c.   a nonsecretor boy.

    d.   What is the probability that their first <u>two</u> children will be nonsecretors of any sex?

Solution

From the data we infer that both parents must be '*Ss*'.

    a. $1.4 \times 1/2 = 1/8$.

    b. $3/4 \times 1/2 = 3/8$.

    c. $3/4 \times 1/2 = 3/8$.

    d. $3/4 \times 3/4 = 9/16$.

6.   Huntington's disease is an autosomal dominant. A woman whose great-grandfather had Huntington's disease marries and is considering having a child. Both her father and grandfather were killed in wars while in their twenties. If she consulted you, how would you advise her about the risks to her child? (Make sure you give her probabilty values for your predictions.)

Solution

Apparently the father and grandfather did not show symptoms of Huntington's disease but since it is a late onset disease they probably would not have expressed it in their early twenties. There is a 1/2 chance that the grandfather inherited the allele, a 1/2 chance that the father received it from him, a further 1/2 chance that the woman received it and then a further 1/2 chance that her future child would receive it. Overall the chance of the woman's child developing the disease is $1/2 \times 1/2 \times 1/2 \times 1/2 = 1/16$.

7.   A man whose mother had cystic fibrosis (autosomal recessive) marries a phenotypically normal woman from outside the family, and the couple consider having a child.

    a.   If the frequency of cystic fibrosis heterozygotes in the general population is 1 in 25, what is the chance that the first child will have cystic fibrosis?

    b.   If the first child does have cystic fibrosis, what is the probability that a second child will be normal?

Solution

    a.   The man must be a heterozygote *Cc*. The probability of his wife being *Cc* is 1/25, and if they are both *Cc* the probability of an affected child is 1/4. Overall the probability of the affected child is $1/25 \times 1/4 = 1/100$.

    b.   The first child shows that both parents must have been *Cc*, so the probability of the next child being *cc* is 1/4.

8.   In peas the gene loci for seed color (alleles *Y* = yellow, *y* = green) and seed shape (alleles *W* = round, *w* = wrinkled) assort independently. A plant of genotype *Yy ww* is crossed to a plant of genotype *yy Ww*. In the progeny, what proportion of peas is expected to be

    a.   yellow round?

    b.   green wrinkled?

Solution

    a.   1/2 will be yellow (*Yy*) and of these 1/2 will be round (*Ww*); overall $1/2 \times 1/2 = 1/4$.

    b.   1/2 will be green (*yy*) and of these 1/2 will be wrinkled (*ww*); overall $1/2 \times 1/2 = 1/4$.

9.   In salmonberry (*Rubus spectabilis*), a pure line with red flowers was crossed to a pure line with white flowers. The F_1 all had red flowers.

    a.   Which phenotype is dominant, red or white?

    b.   If the F_1 is selfed, what proportion of the F_2 is expected to be white? Show all genotypes in your explanation.

Solution

    a.   By definition red is the dominant phenotype.

    b.   If two alleles control this phenotypic difference then the cross was *RR* (red) × *rr* (white), the F_1 was *Rr*, and a likely outcome is that the F_2 will be 3/4 red (*RR*) and 1/4 white (*rr*).

10.   Two black guinea pigs (a male and a female) of the same genotype were allowed to breed and over two years they produced 29 black progeny and 9 brown progeny.

    a.   What were the genotypes of the parental guinea pigs?

    b.   What do you predict would be the types and proportion of progeny from crossing one of the brown guinea pigs back to one of its parents?

    c.   What would be the result of crossing two of the brown guinea pigs?

    d.   Two black offspring were allowed to breed over several years, and they never produced any brown offspring. What possible genotypes could they be? Are there are several possibilities.

Solution

    a.   Let *B* = black and *b* = brown; then the parents must have been *Bb* because the progeny show a phenotypic ratio that is approximately 3:1.

    b.   *bb* × *Bb* -> 1/2 *bb* (brown) and 1/2 *Bb* (black).

    c.   *bb* × *bb* -> all *bb* (brown).

    d.   *BB* × *BB* or *BB* × *Bb*.

11.   A woman is planning to marry her first cousin, but the couple discover that their shared grandfather's sister died in infancy of Tay-Sachs disease (a rare autosomal recessive disorder).

    a.   Draw the relevant parts of the pedigree as described, and show all the genotypes as completely as possible.

    b.   What is the probability that the cousins' first child will have Tay-Sachs disease, assuming that all people who marry into the family are homozygous normal?

Solution

    a.   Pedigree:

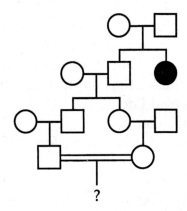

?

    b.   Since the great-grandparents must have both been *Tt*, the grandfather stood a 2/3 chance of being *Tt*. Then there is a 1/2 chance of the allele passing to his children, and then to their children. Finally, if the couple are *Tt* then there is a 1/4 chance of a Tay-Sachs child. Overall the probability is 2/3 × 1/2 × 1/2 × 1/2 × 1/2 × 1/4 = 2/192 = 1/96.

12.   In Drosophila the genes for wing size (*V* = large, *v* = vestigial) and bristle length (*B* = long, *b* = short) are inherited independently. Deduce the genotypes of the parents in the following crosses

	Progeny			
Parents	Large, long	Large, short	Vestigial, long	Vestigial, short
a. large, long × vestig., short	246	0	250	0
b. large, long × vestig., long	302	98	297	101
c. large, long × large, short	150	148	48	51
d. large, short × vestig., long	89	93	90	94

Solution

    a.   *Vv BB × vv bb.*

    b.   *Vv Bb × vv Bb.*

    c.   *Vv Bb × Vv bb.*

    d.   *Vv bb × vv Bb.*

13   A man is brachydactylous (very short fingers; rare autosomal dominant) and his wife is not. Both can taste the chemical phenylthiocarbamide (autosomal dominant; common) but both their mothers could not.

    a.   give the genotypes of the couple.

If the genes assort independently and if they have four children, what is the probability of

    b.   all being brachydactylous?

    c.   none being brachydactylous?

    d.   all being tasters?

    e.   all being nontasters?

    f.   all being brachydactylous tasters?

    g.   none being brachydactylous tatsers?

    h.   at least one being a brachydactylous taster?

    i.   the first child being a brachydactylous tatser?

    j.   the fist two children being brachydactylous?

    k.   having exactly two brachydactylous children?

    l.   the first two children being a brachydactylous taster and a nonbrachydactylous nontaster in any order?

Solution

    a.   If *B* = brachydactylous and *P* = taster, then they are *Bb Pp* (man) × *bb Pp* (woman).

    b. $(1/2)^4 = 1/16.$

    c. $(1/2)^4 = 1/16.$

d. $(3/4)^4 = 81/256$.

e. $(1/4)^4 = 1/256$.

f. $[(1/2)(3/4)]^4 = (3/8)^4 = 81/4096$.

g. $(5/8)^4 = 3125/4096$.

h. $1-(3125/4096) = 971/4096$.

i. $(1/2)(1/4) = 1/8$.

j. $(1/2)(1/2) = 1/4$.

k.   For the first two and only the first two to be brachydactylous, the probability is $(1/2)4 = 1/256$, but the brachydactylous combinations can be $1 + 2 / 1 + 3 / 1 + 4 / 2 +3 / 2 + 4 / 3 + 4$; a total of six combinations making the overall probability $6/256$.

l. $(1/2)(3/4) \times (1/2)(1/4) \times 2 = 3/32$.

C H A P T E R

3

Chromosome Theory of Inheritance

Multiple-Choice Questions

1. Ploidy is
   a. the number of genes in a cell.
   b. the number of chromosomes in a cell.
   c. the amount of DNA in a cell.
   d. the number of chromosomes in a chromosome set.
   e. the number of chromosome sets. •

2. A process that occurs in meiosis but not mitosis is
   a. pairing of homologs. •
   b. chromatid formation.
   c. cell division.
   d. separation of homologous centromeres to opposite poles.
   e. chromosome condensation (shortening).

3. The 'pulling apart' stages of both mitosis and meiosis are called
   a. prophase.
   b. metaphase.
   c. anaphase. •
   d. telophase.
   e. interphase.

4.   A characteristic of homologous chromosomes is that
    a.   they carry alleles for the same genes in the same relative positions.
    b.   they regularly exchange parts by crossing over at meiosis.
    c.   in a diploid cell in interphase they are found in pairs but they do not physically pair.
    d.   they physically pair at meiosis.
    e.   All of the above •

5.   A calico cat becomes pregnant but it is not known which male is the father. There are 6 kittens: 2 calico females, 1 orange female, 1 orange male, and 2 black males. Which of the following best represents the genotype of the male? (Let *R* = orange, *r* = black.)
    a.   *Rr*
    b.   *rr*
    c.   *RR*
    d.   *R* •
    e.   *r*

6.   In humans baldness is caused by a rare allele *B′* that is dominant in males and recessive in females. The only other allele of this gene, *B*, does not cause baldness. A bald man has non-bald parents. What must have been their genotypes? (Mother is written first.)
    a. $BB \times BB′$
    b. $BB′ \times BB′$
    c. $BB′ \times BB$ •
    d. $B′B′ \times BB′$
    e. $BB′ \times B′B′$

7.   In birds the female is the heterogametic sex (ZW) and the male is the homogametic sex (ZZ). It is important for chicken breeders to be able to separate male chicks from female chicks soon after birth, although sexing chickens at this age is extremely difficult. However, a Z-linked dominant allele *B* (barred feathers) can be used to help in this problem because the barred pattern can be identified immediately after hatching. From which cross would all the chicks of one sex be barred and all those of the other sex nonbarred? (*M* = male, *F* = female)
    a.   barred *M* × nonbarred *F*
    b.   nonbarred *M* × barred *F* •
    c.   nonbarred *M* × nonbarred *F*
    d.   barred *M* × barred *F*
    e.   more than one of the above possibilities works

8.   In a plant of 2n = 24, what is the total number of chromatids present during prophase of meiosis?

    a.   6
    b.   12
    c.   24
    d.   48 •
    e.   96

9.   In a plant in which 2n = 24, what is the total number of chromatids present during prophase of mitosis?

    a.   6
    b.   12
    c.   24
    d.   48 •
    e.   96

10.   In a plant in which 2n = 24, how many bivalents should be visible at metaphase I of meiosis?

    a.   6
    b.   12 •
    c.   24
    d.   48
    e.   96

11.   Which diagram most accurately shows the arrangement of homologous chromosomes during the first metaphase of meiosis?

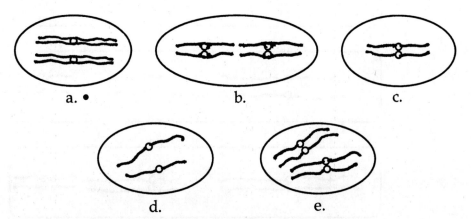

    a. •   b.   c.

             d.   e.

12. Which diagram most accurately shows the arrangement of homologous chromosomes during the metaphase of mitosis?

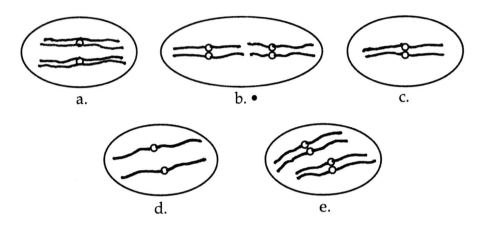

a.   b. •   c.

d.   e.

13. In a diploid organism 2n = 6, and there are two long, two intermediate, and two short chromosomes. What is the most accurate representation of a gamete resulting from meiosis in this organism?

a. •

b.

c.

d.

e.

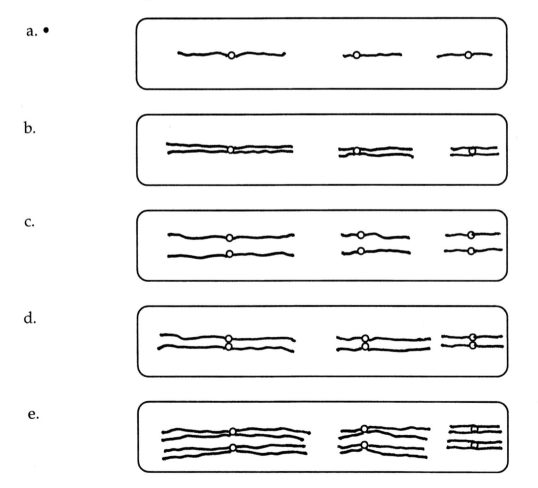

14. In a dihybrid the genes are located on different chromosomes. Which of the following diagrams does *not* represent a stage of meiosis in this organism?

a.

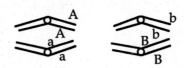

b.

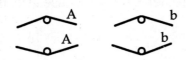

c.

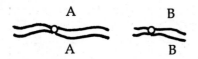

d. •

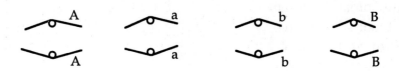

e.

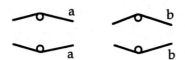

15. A woman whose father was a hemophiliac is expecting a baby and fetal tests show that it is a boy. What is the likelihood that the boy will be a hemophiliac?
a. 100%
b. 75%
c. 50% •
d. 25%
e. 0%

16. A man with achondroplasia (rare autosomal dominant) marries a woman of normal height. Both have normal color vision but the wife's father was red-green colorblind (rare X-linked recessive). What is the probability that their first child will be colorblind and of normal height?
a. 0
b. 1/2
c. 1/4
d. 1/8 •
e. 1/16

17. What proportion of human sperm will have centromeres all of which are from the man's father?
   a. $(1/2)^{23}$ •
   b. $(1/4)^{46}$
   c.   1/2
   d.   All of them
   e.   None of them

18. The following pedigree concerns a rare hereditary disease affecting muscles.

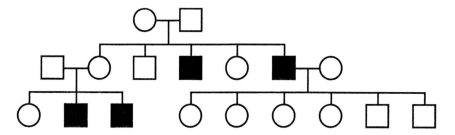

What is the most likely mode of inheritance of this disease?
   a.   autosomal dominant
   b.   autosomal recessive
   c.   X-linked dominant
   d.   X-linked recessive •
   e.   Y-linked

19. The following pedigree shows the inheritance of attached earlobes (black symbols) and unattached earlobes (white symbols). Both alternative phenotypes are quite common in human populations.

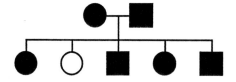

If the alternative phenotypes are determined by alleles of one gene, then attached earlobes
   a.   must be autosomal dominant. •
   b.   must be autosomal recessive.
   c.   must be X-linked dominant.
   d.   must be X-linked recessive.
   e.   could be more than one of the above.

20. The following pedigree is for a rare hereditary disease of the kidneys.

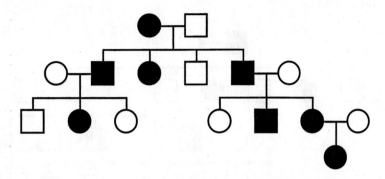

Which of the following modes of inheritance is <u>most likely</u> for this disease?
    a.   autosomal dominant •
    b.   autosomal recessive
    c.   X-linked dominant
    d.   X-linked recessive
    e.   cannot tell from this pedigree

21. In *Drosophila* the recessive allele for white eyes is X-linked, and the recessive allele for vestigial wings is autosomal. A female heterozygous for both these alleles is crossed to a white-eyed vestigial-winged male. What proportion of daughters will be white-eyed with vestigial wings?
    a.   100%
    b.   75%
    c.   50 %
    d.   25% •
    e.   0%

Open-Ended Questions

1. Consider a pea plant of genotype *Aa Bb* (the *Aa* and *Bb* allele pairs are on different chromosome pairs). Draw simple meiosis diagrams to show how the four gamete types *AB*, *Ab*, *aB*, and *ab* are produced in equal frequency.

Solution
See Figure 3-30 in text.

2.   The following pedigree concerns a rare condition involving involuntary movements of the eyeballs.

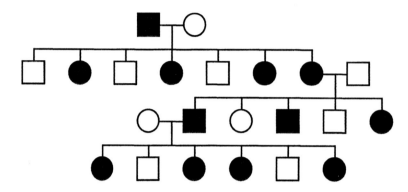

Determine the most likely mode of inheritance of this phenotype and write the genotypes of as many members of the pedigree as possible.

Solution

Phenotype found in every generation; males transmit only to females; females transmit to equal numbers of both sexes; therefore most likely X-linked dominant. All affected females heterozygous $X^D X^d$; all affected males $X^D Y$.

3.   A man who has brachydactyly (short fingers, inherited as a rare autosomal dominant) marries a woman who has normal hands and is red-green colorblind (inherited as an X-linked recessive). List the genotypic and phenotypic proportion expected in their sons and their daughters.

Solution

The man is $Bb\ X^C Y$ the woman is $bb\ X^c X^c$

The sons will be	$1/2\ Bb\ X^c$	brachydactyly, colorblind, and
	$1/2\ bb\ X^c$	colorblind.
The daughters will be	$1/2\ Bb\ X^C X^c$	brachydactyly, and
	$1/2\ bb\ X^C X^c$	normal.

4.   A couple with normal vision have a daughter with red-green colorblindness (rare X-linked recessive). The man sues his wife for divorce on the grounds of infidelity saying it is impossible that he fathered the child. The man's lawyers call you in as a genetics advisorwhat advice do you give?

Solution

For a woman to be homozygous for an X-linked recessive, <u>both</u> parents must transmit the recessive allele. The wife might be heterozygous, but the man would be hemizygous for the colorblind allele but he is obviously not colorblind so this is impossible. It is remotely possible (but unlikely) that a mutation to the recessive allele occurred in the man's sperm. However it is much more likely that another man fathered the child.

5.   A young woman has just got married but she is worried about having a child because her mother's only sister had a son with Duchenne muscular dystrophy (DMD). The young woman has no brothers or sisters. (DMD is a rare X-linked recessive disorder.)

  a.   Draw the relevant parts of the pedigree of the family described above. (Be sure to include the grandmother, the three women mentioned, and all their mates.)

  b.   State the most likely genotype of everyone in the pedigree.

  c.   Calculate the probability that the young woman's first child will have DMD.

Solution

  a.   Pedigree:

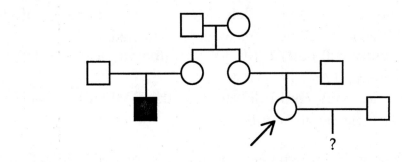

  b.   *D   Dd*
          D   *Dd   D-*   *D*
                d   *D-*   *D*

  c.   The grandmother must have been *Dd*. There is a 1/2 chance that the mother is *Dd*, and if so a further 1/2 chance that the woman herself is *Dd*. If she is, 1/2 of her sons will have DMD. Since the probability of a son is also 1/2, the overall probability is $1/2 \times 1/2 \times 1/2 \times 1/2 = 1/16$.

6.   Mary and her brother Paul are both healthy adults. Mary and her husband Mike have a healthy baby girl and Mary is pregnant again. They learn that Mary and Paul's mother has just had a baby by a second marriage, and the baby has Duchenne muscular dystrophy (DMD). (DMD is a rare X-linked recessive disorder.)

  a.   Draw the pedigree of the individuals described and relevant relatives.

  b.   Do Mary and Mike have to worry about their new baby having DMD? Explain.

  c.   If so, what is the probability the baby will have DMD?

  d   Why didn't Mary and Paul and the first baby get DMD?

Solution

    a.   Pedigree:

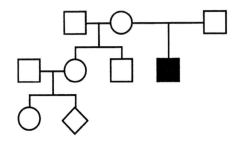

    b.   The mother must be a heterozygote X^DX^d. Paul is unaffected so he must be X^DY and can not pass on the *d* allele. Mary has a 1/2 chance of also being a heterozygote X^DX^d; if so half her sons would be affected.

    c.   The probability will be 1/2 (Mary) × 1/2 (having a son) × 1/2 (son having DMD allele *d*. Overall, 1/8.

    d.   Mary and her baby could not express the DMD allele because they are female. Paul was lucky not to have been X^dY.

7.   The diagram below shows two chromosome pairs of a diploid cell in interphase. Two heterozygous allele pairs are also shown, *A/a* and *B/b*. With the aid of diagrams, show how this cell would go through meiosis and mitosis. (Illustrate as many stages as you think necessary to show the main features of the processes.) Make sure you show clearly the meiotic and mitotic products with their genotypes.

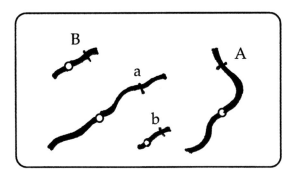

Solution
See figures in chapter.

8.   The following pedigree is for a rare disease. Could the pedigree be explained *in theory* by

    a.   autosomal recessive inheritance?

    b.   autosomal dominant inheritance?

    c.   X-linked recessive inheritance?

    d.   X-linked dominant inheritance?

    e.   Y-linkage?

    f.   Which is most likely?

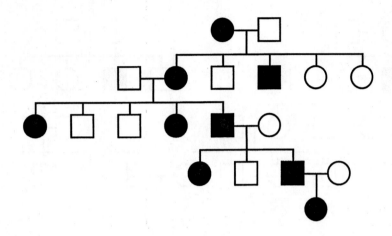

Solution

    a.   Possible if heterozygotes marry into family.

    b.   Possible.

    c.   Impossible because unaffected males produce affected daughters.

    d.   Impossible because there is male to male transmission.

    e.   Impossible because there are affected females.

    f.   b is most likely as it requires only one dominant allele passing down through the generations.

9.   In mice, dwarfism is caused by an X-linked recessive allele, and pink coat is caused by an autosomal dominant allele (coat color is normally brown). If a dwarf female from a pure line is crossed to a pink male from a pure line, what will be the phenotypic ratios in the F_1 and F_2 in each sex? (Invent and define your own gene symbols.)

Solution

female: $PP\ X^d X^d$; male $pp\ X^D Y$.

F_1:   Females $Pp\ X^D X^d$; males $Pp\ X^d Y$

F_2:   Females   3/8 *P- Dd*   brown large,

                          3/8 *P- dd*   brown dwarf,

                          1/8 *pp Dd*   pink large,

                          1/8 *pp dd*   pink dwarf.

         Males   3/8 *P- D Y*   brown large,

                          3/8 *P- d Y*   brown dwarf,

                          1/8 *pp D Y*   pink large,

                          1/8 *pp d Y*   pink dwarf.

10.   A married couple discovers that there was a case of the X-linked recessive disorder Duchenne Muscular Dystrophy (DMD) in the grandparental generation on <u>both</u> sides of their family. The pedigree is as follows:

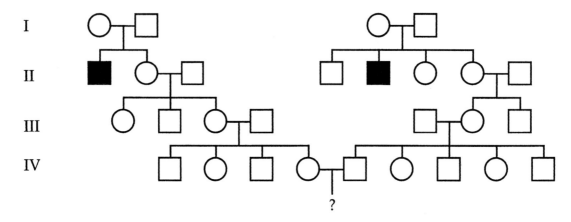

a. Should the DMD case on the man's side affect the anxiety level of the couple at all? Explain.

b. What is the probability that their first child will be a boy with DMD?

Solution

a. No because the man apparently does not have the gene so he cannot transmit it.

b. In the left branch in generation I the woman is a carrier, so there is a $1/2 \times 1/2 \times 1/2$ chance that the woman in generation IV will be heterozygous, and then $1/2$ of her children will be sons and $1/2$ of them will have DMD. Overall, $(1/2)^5 = 1/32$.

11. In *Drosophila* the recessive allele *v* causes vermilion eyes (wild type is red). At a different gene the allele *b* causes black body (wild type is brown). A vermilion brown female was crossed to a red, black male and the F_1 was found to be equal numbers of completely wild-type (red, brown) females and vermilion, brown males. The F_2 was as follows

Females:
  72 vermilion brown,
  23 vermilion black,
  70 red brown,
  24 red black.
Males:
  73 vermilion brown,
  24 vermilion black,
  76 red brown,
  25 red black.

Explain the inheritance of these colors and show genotypes of parents, F_1 and F_2.

Solution

The F_1 shows that brown is dominant to black, apparently in a standard autosomal manner. However males and females differ for the other pair of phenotypes suggesting that they are X-linked.

The F_2 shows a 3:1 ratio of brown to black, confirming phenotypic determination by a single pair of autosomal alleles. There is a 1:1 ratio of vermillion to red in both sexes, suggesting that the F_1 cross must have been $\qquad$ $X^V X^v\ Bb \quad \times \quad X^v Y\ Bb$

and the parents $\qquad\qquad\qquad\qquad\qquad\qquad$ $X^v X^v\ BB \quad$ and $X^V X^V\ bb$

The F_2 shows a 3/4:1/4 ratio for the autosomal gene and 1/2:1/2 ratio for the X-linked gene in both sexes: in combination these give a 3/8:3/8:1/8:1/8 ratio in each sex.

12.   The *P*, *Q*, and *R* loci are all on different autosomes in a certain mammalian species. The *S* locus however is sex linked. The following cross was made

   Female *Pp Qq Rr Ss* $\qquad \times \qquad$ Male *pp Qq rr S*

With respect to these four genes
   a.   What proportion of sons will phenotypically resemble their fathers?
   b.   What proportion of sons will phenotypically resemble their mothers?
   c.   What proportion of daughters will phenotypically resemble their mothers?
   d.   What proportion of daughters will phenotypically rsemble their fathers?
   e.   What proportion of daughters will genotypically resemble their mothers?
   f.   What proportion of progeny will have one of the the parental phenotypes?

Solution

	S	P	Q	R	Total
a.	1/2	1/2	3/4	1/2	3/32 of sons
b.	1/2	1/2	3/4	1/2	3/32 of sons
c.	1	1/2	3/4	1/2	3/16 of daughters
d.	1	1/2	3/4	1/2	3/16 of daughters
e.	1/2	1/2	1/2	1/2	1/16 of daughters
f.	6/32 of sons = 3/32 of total; 6/16 of daughters = 6/32 of total				

Therefore a total of 9/32 will resemble a parent.

CHAPTER

4

Extensions of Mendelian Analysis

Multiple-Choice Questions

1. In a certain breed of dog, the alleles B and b determine black and brown coats respectively. However, the allele Q of a gene on a separate chromosome is epistatic to the B and b color alleles resulting in a grey coat. (q has no effect on color.) If animals of genotype $Bb\,Qq$ are intercrossed, what phenotypic ratio is expected in the progeny?
   a. 9 grey, 3 brown, 4 black
   b. 1 black, 2 grey, 1 brown
   c. 9 black, 6 brown, 1 grey
   d. 9 black, 4 grey, 3 brown
   e. 12 grey, 3 black, 1 brown•

2. Two rabbits with chinchilla and himalayan coats are intercrossed. Their progeny are 1/2 chinchilla, 1/4 himalayan, and 1/4 albino. The parental genotypes were
   a. $c^{ch}c^{ch} \times c^h c^{h.}$
   b. $c^{ch}c^h \times c^h c.$
   c. $c^{ch}c \times c^h c. \bullet$
   d. $c^{ch}c^h \times cc.$
   e. $c^{ch}c^{ch} \times c^h c.$

3. In mice the enzyme esterase shows two forms, fast and slow, when studied electrophoretically. A mouse from a pure line with only the fast form is crossed to a mouse from a pure line with only the slow form. The F_1 all show both forms and the F_2 shows 1/4 fast, 1/4 slow, and 1/2 with both. This is an example of
    a.   incomplete dominance.
    b.   complementation.
    c.   epistasis.
    d.   codominance. •
    e.   incomplete penetrance.

4. In sweet peas the allele *C* is needed for color expression (*c* results in no color, that is, white). The precise color expressed is determined by the alleles *R* (red) and *r* (blue). A cross between certain red and blue plants resulted in progeny as follows:
    3/8 red
    3/8 blue
    1/4 white
    What were the genotypes of the plants crossed?
    a.   *CC RR* × *Cc Rr*
    b.   *Cc rr* × *CC Rr*
    c.   *Cc Rr* × *Cc Rr*
    d.   *Cc Rr* × *Cc rr*•
    e.   *cc Rr* × *Cc Rr*

5. In *Drosophila* the eye color peach is determined by an autosomal recessive allele *p*. On another chromosome the autosomal recessive *s* suppresses peach restoring the wild type red eye color. When two pure-breeding red strains 1 and 2 are intercrossed the F_1 is also red. However when the F_1 is backcrossed to strain 1 the progeny show 3/4 red and 1/4 peach flies. The genotype of the strain 1 must have been
    a.   *PP SS*
    b.   *Pp ss*
    c.   *pp SS*
    d.   *Pp Ss*
    e.   *pp ss* •

6. In mice the allele for color expression is *C* (*c* = albino). Another gene determines color (*B* = black and *b* = brown). Yet another gene modifies the amount of color so that *D* = normal amount of color and *d* = dilute (milky) color. Two mice that are *Cc Bb Dd* are mated. What proportion of progeny will be dilute brown?
    a.   9/16
    b.   3/64 •
    c.   9/64
    d.   1/4
    e.   3/16

7.   Drosophila eyes are normally red. Several purple-eyed strains have been isolated as spontaneous variants (mutants) and the purple phenotype has been shown to be inherited as a Mendelian autosomal recessive in each case. To investigate allelism between these different purple mutations, two purple-eyed pure strains were crossed. If the purple mutations are in different genes (that is, they are *not* allelic), the F_1 is expected to be

    a.   all red. •
    b.   all purple.
    c.   3/4 red, 1/4 purple.
    d.   1/2 red, 1/2 purple.
    e.   3/4 purple, 1/4 red.

8.   *Drosophila* eyes are normally red. Several purple-eyed strains have been isolated as spontaneous variants (mutants) and the purple phenotype has been shown to be inherited as a Mendelian autosomal recessive in each case. To investigate allelism between these different purple mutations, two purple-eyed pure strains were crossed. If the purple mutations are in the same gene (that is, they *are* allelic), the F_1 is expected to be

    a.   all red.
    b.   all purple. •
    c.   3/4 red, 1/4 purple.
    d.   1/2 red, 1/2 purple.
    e.   3/4 purple, 1/4 red.

9.   In the haploid fungus *Neurospora* the mycelium is normally pink. Two albino strains were obtained in different labs and intercrossed. The progeny were 3/4 albino and 1/4 pink. The two parental strains can best be represented as

    a.   al-1$^+$ al-2$^+$ × al-1 al-2.
    b.   al-1$^+$ al-2 × al-1 al-2$^{+.}$ •
    c.   al-1^a × al-1$^{b.}$
    d.   al$^+$ al × al$^+$ al.
    e.   al × al.

10.   In *Drosophila* the recessive alleles for brown and scarlet eyes (of two independent genes) interact so that *bb ss* is white. If a pure-breeding brown is crossed to a pure-breeding scarlet, what proportion of the F_2 will be white?

    a.   3/4
    b.   7/16
    c.   1/4
    d.   1/16 •
    e.   13/16

11. In chickens the dominant allele Cr produces the 'creeper' phenotype (having extremely short legs). However, the creeper allele is lethal in homozygous condition. If two creepers are mated, what proportion of the progeny will be creepers?
   a. 1/4
   b. 1/2
   c. 3/4
   d. 1/3
   e. 2/3 •

12. The mycelium of haploid fungi will readily fuse with another of the same species. If the fusing strains are genetically different, the fusion product is called a heterokaryon, two different haplid nuclei in a common cytoplasm. In *Neurospora*, the nuclei do not fuse and remain separate. Two different *Neurospora* non-wild type strains were adenine-requiring, that is, they required adenine to be added to the medium in order for growth to occur. These strains were fused as a heterokaryon. The heterokaryon did not require adenine to be added to the medium and would grow well without it, just like wild-type strains. This result most likely means:
   a. the alleles determining adenine-requirement were of the same gene.
   b. the alleles determining adenine-requirement were of different genes. •
   c. a suppressor allele arose spontaneously.
   d. independent assortment had produced a wild-type genotype.
   e. the two fusing strains each contained a few wild-type nuclei.

13. In one species of beetles the number of spots on the wing covers is determined by alleles of one gene. Two pure lines with 2 spots and 3 spots respectively were intercrossed and the F_1 all had 3 spots. If an F_2 is produced, what phenotypic ratio is predicted?
   a. all have 3 spots
   b. all have 2 spots
   c. 3/4 have 3 spots, 1/4 have 2 spots •
   d. 1/2 have 2 spots, 1/2 have 3 spots
   e. 9/16 have 3 spots, 7/16 have 2 spots.

14. In swine when a pure-breeding red is crossed to a pure-breeding white the F_1 are all red. However, the F_2 shows 9/16 red, 1/16 white, and 6/16 are a new color sandy. The sandy phenotype is most likely determined by
   a. complementary alleles of two different genes.
   b. a heterozygote of the alleles determining red and white.
   c. recessive epistasis.
   d. dominant epistasis.
   e. the presence of a dominant allele at either of two loci that both must carry dominant alleles to determine red. •

15. In Andalusian fowls BB = black, and bb = white but the heterozygote is blue. If two heterozygotes are mated, what proportion of their offspring will be blue?
   a. 3/4
   b. 2/3
   c. 1/4
   d. 1/2 •
   e. 9/16

Open-Ended Questions

1. Consider five different F_1 dihybrids that when selfed produce F_2 phenotypic ratios that are
   a. 9:7
   b. 12:3:1
   c. 9:3:4
   d. 15:1
   e. 13:3

If those F_1 individuals are each testcrossed, what phenotypic ratio would result in each case?

Solution
   a. 1:3
   b. 2:1:1
   c. 1:1:2
   d. 3:1
   e. 3:1

2. Four babies were born in a hospital on a night in which an electrical blackout occurred. In the confusion that followed, their identification bracelets were mixed up. The blood groups of the babies were O, A, B, and AB. The four pairs of anxious parents were O and O, AB and O, A and B, and B and B. You are called in as a genetic advisor to determine which baby belongs to which parents. What is your analysis?

Solution
The O baby must come from the O × O mating. The only mating that can produce AB is the A × B mating. From the remainder the AB × O mating is the only one that can produce A, leaving the B × B mating as the parents of the B child.

3. In dogs, the Mexican hairless phenotype never breeds true. When Mexican hairless are interbred the progeny are 2/3 hairless and 1/3 normal. It is also observed that some deformed puppies are born dead. Propose a genetic explanation for these observations.

Solution
H produces hairlessness, but is lethal (or sublethal) in homozygous state. So all matings are $Hh \times Hh$, and of the survivors 2/3 are *Hh*.

4.   A mouse from a pure line with the recessive phenotype bent tail is crossed to a mouse from a pure line with the recessive phenotype dwarf. The F_1 were all wild-type in appearance. The F_2 showed a ratio of 9/15 wild-type:3/15 bent tail:3/15 dwarf. Propose a genetic explanation for this unusual phenotypic ratio in the F_2, giving genotypes to parents, F_1 and F_2.

Solution
The double mutant genotype *bb dd* must be lethal. F_1 was *Bb Dd*, which gave 9:3:3:1 on selfing; the 1 component is lethal so will be missing.

5.   In mice agouti is determined by the allele *A* and nonagouti is *a*; black and brown are *B* and *b*; and spotted and unspotted are *S* and *s*. If trihybrid mice are intercrossed, what proportion of solid (that is, nonagouti) back spotted phenotype is expected in the progeny assuming independent assortment?

Solution
1/4 solid, 3/4 black, 3/4 solid; overall $1/4 \times 3/4 \times 3/4 = 9/64$.

6.   In corn the recessive allele albescent (*al*) inhibits chlorophyll development. A green plant produces 3/4 green seedlings and 1/4 white seedlings which soon die.
    a.   What must have been the genotype of the green parent plant?
    b.   What was the genotype of the white seedlings and why did they die and how could they grow at all?
    c.   Can *al* be considered a lethal allele? Explain.
    d.   The recessive allele *vr* (virescent) causes the seedling to start off white but green color develops slowly. Would you expect this to be a lethal allele?

Solution
    a.   *Al al*
    b.   *al al*. They grew to be seedlings using maternal food laid down in the seed; once this was exhausted they died because they could not photosynthesize.
    c.   Yes; will die as suggested in *b*.
    d.   Depends on timing; if chlorophyll develops before food exhaustion, they will be fine.

7.   In the haploid fungus *Neurospora* the albino allele (*al*) is on chromosome 1 and yellow (*ylo*) is on chromosome 6. A cross was made between an *al* and a *ylo* strain and the progeny were
    1/2 albino
    1/4 yellow
    1/4 wild type (pink)

a. Explain this result in terms of epistasis.
b. Show genotypes of all strains.
c. Speculate on the sequence in which these genes act in a biochemical pathway.

Solution

a. *al* must be epistatic to *ylo*, so that they are both white.
Progeny:

+ +	pink
al +	white
al ylo	white
+ *ylo*	yellow

b. see above
c. al enzyme must act first, converting colorless compound to yellow; then second enzyme converts yellow to pink

$$\text{white} \xrightarrow{\quad al \quad} \text{yellow} \xrightarrow{\qquad ylo \qquad} \text{pink}$$

8. In gerbils, crosses were made between animals with coat colors called orange, beige, and cream (the animals were not necessarily pure-breeding).

beige × beige -> 3/4 beige, 1/4 cream

beige × cream -> 1/2 beige, 1/2 orange

orange × cream -> 1/2 orange, 1/2 cream

beige × beige -> 3/4 beige, 1/4 orange

orange × orange -> 3/4 orange, 1/4 cream

Invent a unifying hypothesis to explain these results, and show all genotypes under your scheme.

Solution

All are single gene ratios, so it is most likely a case of multiple alleleism.

Cross 1 shows beige > cream.
Cross 3 shows beige > orange.
Cross 4 shows orange > cream.

Therefore, dominance series is $b > o > c$.

9. Two pure-breeding lines of albino poodles were intercrossed and the F_1 was always found to be black. If F_1 animals were intercrossed the F_2 was always 9/16 black and 7/16 white.

a. Propose a genetic explanation showing genotypes of the parental lines, F_1 and F_2.

b. If the F_1 is backcrossed to each parent, what phenotypic ratios are expected?

c. One of the F_2 black animals was crossed with one F_2 white, and over the years the cross produced 25% black And 75% white poodles. What were the genotypes of the F_2 animals crossed?

Solution

 a. *AA bb* and *aa BB* are albino; *A- B-* is black.

AA bb × *aa BB* —> *Aa Bb* —> 9 *A- B-*: 3 *A- bb*; 3 *aa B-*; 1 *aa bb*

 b. *Aa Bb* × *AAbb* —> 1/2 *A- Bb* (black); 1/2 *A- bb* (albino). Same ratio in other backcross *Aa Bb* × *aa BB*.

 c. Cross is *A- B-* × *aa bb*; to get 25% black the black parent must have been dihybrid *Aa Bb*.

Aa BB × *aa bb* —> 1/4 *Aa Bb* (black); 3/4 rest (albino).

10. Achondroplastic dwarfism is a rare autosomal dominant phenotype in humans.

 a. On very rare occasions two achondroplastic dwarfs marry. What would be the most likely genotypes of these two individuals who marry? Explain your answer and any gene symbols that you use.

 b. Based on simple Mendelian rules, what genotypic and phenotypic ratios are expected in the offspring of such marriages?

 c. In reality the progeny ratio in the children of such marriages is close to 2/3 dwarfs and 1/3 normal height. Is this a departure from Mendelian expectations? If so suggest an explanation.

Solution

 a. *Aa*; because *A* is rare so *AA* is very unlikely.

 b. 1/4 *AA*, 1/2 *Aa* (dwarf) and 1/4 *aa* (normal height).

 c. Yes, *AA* is probably lethal leaving 2/3 *Aa* and 1/3 *aa*.

11. In Cocker Spaniel dogs, lines are available with the coat colors bronze, cream, golden, and silver. When any pure-breeding silvers are crossed with pure-breeding creams from one particular line, the F_1 puppies are all cream. When the cream F_1 dogs are interbred, the resulting F_2's are

 12/16 cream
 2/16   bronze
 1/16   silver
 1/16   golden

 a. Formulate a model to explain the genetic determination of these four coat colors. (Invent your own allele symbols and define them clearly.)

 b. Show genotypes of parental, F_1 and F_2 animals.

Solution

 a. Cream:noncream is 3:1 so must be dominant pistasis over other colors. Silver:bronze:gold is 1:2:1 so this must be incomplete dominance.

 b. Overall *cc ss* (silver) × *CC SS* (cream)—> *Cc Ss* (cream)

—> 9/16 *C- —-*: 1/16 *cc ss* (silver): 2/16 *cc Ss* (bronze): 1/16 *cc SS* (gold).

12.   A man with extra digits on his hands and feet (polydactyly, an autosomal dominant) marries an unaffected woman. Their first child has polydactyly but the second child's hands and feet appear normal. The second child married a woman with normal hands and feet but the first two children showed polydactyly. Propose a genetic explanation for this unusual result.

Solution
The unaffected child did carry the polydactyly allele but it was not expressed, whereas when passed on to the next generation it was expressed. Clearly this allele shows variable penetrance.

13.   Summer squash may be long, round, or oval in shape. The color may be green, white, or yellow. A long white variety was crossed with a round green one and an oval yellow F_1 resulted. The F_2 showed the following phenotypes

    18 long green
    34 long yellow
    16 long white
    36 oval green
    68 oval yellow
    32 oval white
    17 round green
    32 round yellow
    15 round white
    a.   Provide a genetic explanation of these results. Be sure to define your genotypes, and show the constitution of parents, F_1 and F_2.
    b.   Predict the genotypic and phenotypic proportions in the progeny of a cross between long yellow and round yellow.

Solution
    a.   The long:oval:round phenotypes show a 1:2:1 ratio reflecting incomplete dominance (oval heterozygous). The same is true for green:yellow:white, once again showing incomplete dominance (yellow heterozygous). Therefore 9 phenotypes are expected in the ratio 1:2:1:2:4:2:1:2:1 as observed.
    b.   The cross must be *LL Gg × ll Gg*, so the progeny are expected to be

1/4 *Ll GG* (oval green)
1/2 *Ll Gg* (oval yellow)
1/4 *Ll gg* (oval white)

14.   In the stems of corn seedlings purple color (anthocyanin) is determined by *C* and colorless (white) by *c*. At an unlinked locus an allele *K* prevents the expression of *C*, whereas *k* does not. If a true-breeding colored variety is crossed with a white one of genotype *ccKK*
    a.   what genotypic and phenotypic proportions are predicted in the F_1 and F_2.
    b.   At a third unlinked locus, a pair of alleles determines the specific color, *R* giving red and *r* giving yellow. If in the above cross the colored variety was *RR* and the white was *rr*, what would be the phenotypic ratio in the F_2?

Solution

a.   *CC kk × cc KK* —> *Cc Kk* —> 3/16 C- *kk* (colored):13/16 rest (white)

b.   *CC kk RR × cc KK rr* — *Cc Kk Rr* —> 13/16 white

$$3/16 \text{ color} \to 3/4 \text{ red}$$
$$\to 1/4 \text{ yellow}$$

Overall 52/64 white, 9/64 red, 3/64 yellow.

15.   In many plants a 'monster' flower phenotype is commonly found called peloria, consisting of several fused flowers. In Antirrhinum two pure lines of the following phenotypes were crossed

peloria red × normal white

The F_1 individuals were all pink normal, and the F_2 was as follows

3/16 normal red
6/16 normal pink
3/16 normal white
1/16 peloric red
2/16 peloric pink
1/16 peloric white

Provide a genetic explanation for these results giving genotypes of parental strains, F_1 and F_2.

Solution

Clearly normal flower form (*P*) is dominant over peloria (*p*), and red (*R*) is incompletely dominant over white (*r*) so that *Rr* is pink. The F_2 is a 3:1 ratio for normal:peloria, and a 1:2:1 ratio for red:pink:white as expected.

16.   A pure line of wheat with red seeds is crossed to a pure line with white seeds. The F_1 was all red-seeded but the F_2 showed a ratio of 15 red:1 white.

a.   Explain this 15:1 ratio genetically.

b.   One of the red F_2 plants was backcrossed to the white parent and the progeny were all red. When these red progeny were intercrossed or selfed a ratio of 3 red:1 white was observed in the next generation. Is this result compatible with those described above? Explain.

Solution

a.   The ratio in the F_2 suggests duplicate genes: either *A* or *B* or both can cause redness whereas *aa bb* is white.

b.   The red F_2 individual must have been homozygous for either *AA* or *BB* or both. However, the offspring of the backcross are obviously monohybrid because they segregate in a 3:1 ratio on selfing. So, they must be either *Aa bb* or *aa Bb*, and the red F_2 must have been either *AA bb* or *aa BB*.

17. In rats, the *A* and *R* loci interact as follows
    A- R- = grey
    A- rr = yellow
    aa R- = black
    aa rr = cream

At a third locus, the recessive allele *c* is epistatic to both the *A* and *R* systems and results in albino.

Four different homozygous albino lines, each crossed with a true-breeding grey strain produced grey F_1s, which in turn produced the F_2s shown in the table.

F_2 phenotypes

Albino line	grey	yellow	black	cream	albino
1	174	0	65	0	80
2	48	0	0	0	16
3	104	33	0	0	44
4	292	87	88	32	171

   a.   Determine the genotype of each albino line
   b.   Show how the observed ratios are produced.

Solution

   a.   Ratios of colors are either 3:1 (one allelic difference) or 1:0 (no difference) or 9:3:3:1 (two allelic differences).
   1 *aa RR cc*
   2 *AA RR cc*
   3 *AA rr cc*
   4 *aa rr cc*

18. In mice, wild types have an almost black coat. Geneticists have developed many pure lines of color variants. Two such pure lines are blue and grey. These lines were used in crosses with the following results

Cross parents	F_1	F_2
1 wild × blue	wild	18 wild, 5 blue
2 wild × grey	wild	27 wild, 10 grey
3 blue × grey	wild	133 wild
		41 blue
		46 grey
		17 milky

   a.   Propose a genetic explanation for these results.
   b.   Predict the F_1 and F_2 phenotypic ratios from crossing milky with blue and grey pure lines.

Solution

    a. $1 ++ ++ \times ++ bb ->++ +b -> 3 ++ +-:1 ++ bb$

           $2 ++ ++ \times gg ++ -> +g ++ -> 3 +- ++:1 gg ++$

           $3 ++ bb \times gg ++ -> +g +b ->$ $9 +- +-$ wild

                                                 $3 +- bb$ blue

                                                 $3 gg +-$ grey

                                                 $1 gg bb$ milky

    b. $gg\ bb \times ++ bb -> F_1 -> 3/4 +- bb$ (blue)$:1/4\ gg\ bb$ (milky)

           $gg\ bb \times gg\ ++ -> F_1 -> 3/4\ gg +-$ (grey)$:1/4\ gg\ bb$ (milky)

19.   In mice, the alleles B and b cause black and brown coat color respectively. However, the recessive allele g of an unlinked gene prevents the expression of both B and b, so animals of the constitution gg are always gray in color no matter what their B or b constitution is (for example $BBgg$ is gray not black). The allele G has no effect on color. The following crosses were made:

    Cross 1   gray $\times$ black

    Progeny   1/2 gray

                       3/8 black

                       1/8 brown

    Cross 2   black $\times$ brown

    Progeny   1/4 gray

                       3/8 black

                       3/8 brown

    a.   Show the genotypes of the parents and progeny in cross 1.

    b.   Show the genotypes of parents and progeny in cross 2.

(Be sure to show your reasoning.)

Solution

    Cross 1 gray:color is 1:1, and black:brown is 3:1

    $gg\ Bb \times Gg\ Bb$

    Cross 2 color:gray is 3:1, and black to brown is 1:1

    $Gg\ Bb \times Gg\ bb$

20.   *Nasturtiums* are normally orange. Two pure yellow lines were obtained from different sources and crossed as follows.

    yellow 1 $\times$ orange -> F_1 orange -> F_2 3/4 orange, 1/4 yellow

    yellow 2 $\times$ orange -> F_1 orange -> F_2 3/4 orange, 1/4 yellow

    yellow 1 $\times$ yellow 2 -> F_1 orange -> F_2 9/16 orange, 7/16 yellow

Deduce the full genotypes of parents, F_1 and F_2 in all crosses.

Solution

y1 y1 ++ × ++ ++ —> +y1 ++ —> 3/4 +- ++:1/4 y1 y1 ++

+ + y2 y2 × ++ ++ —> ++ +y2 —> 3/4 ++ +-:1/4 ++ y2 y2

y1 y1 ++ × ++ y2 y2 —> +y1 +y2 —>   *9 +- +-*

                                                 3 y1 y1 +-

                                                 3 +- y2 y2

                                                 1 y1 y1 y2 y2

21. In the leopard frog, *Rana pipiens*, two pure lines with abnormal dorsal pigmentation phenotypes were isolated, 'kandiyohi' and 'burnsi' (see drawings). They were analyzed genetically by making various crosses as shown in the table, which also shows the phenotypic ratios obtained. Note that in one cross another phenotype appeared, 'mottled'.

Cross	Parents	F_1	F_2
1	kandiyohi × wild	kandiyohi	3 kandiyohi:1 wild
2	burnsi × wild	burnsi	3 burnsi:1 wild
3	F_1 kandiyohi × F_1 burnsi	1 wild type	
		1 kandiyohi	
		1 burnsi	
		1 mottled	
4	F_1 mottled × wild type	1 wild type	
		1 kandiyohi	
		1 burnsi	
		1 mottled	

Using gene symbols of your own choosing, show
    a.   the genetic determination of the four phenotypes
    b.   the genotypes of all individuals in the four crosses.

Solution

    a., b.   1 *KK ++ × ++ ++ —> K+ ++*

              2 *++ BB × ++ ++ —> ++ B+*

              3 *K+ ++ × ++ B+ —>*   *++ ++*

                                        K+ ++

                                        ++ B+

                                        K+ B+

              4 *K+ B+ × ++ ++ —>*   *++ ++*

                                        K+ ++

                                        ++ B+

                                        K+ B+

5

Linkage I: Basic Eukaryotic Chromosome Mapping

Multiple-Choice Questions

1. A plant of genotype *CC dd* is crossed to *cc DD* and an F_1 testcrossed to *cc dd*. If the genes are unlinked the percentage of *CC DD* recombinants will be
   a. 10
   b. 20
   c. 25 •
   d. 50
   e. 75

2. In *Drosophila*, the genes *R* and *S* are linked. Flies of genotypes *RR ss* and *rr SS* are crossed and an F_1 obtained. The F_1 allele arrangement is called
   a. coupling.
   b. repulsion. •
   c. complementary.
   d. recombinant.
   e. crossover.

3.   In maize, the genes *W* and *D* are so tightly linked that virtually no crossovers occur between them. A dihybrid *Wd/wD* is testcrossed to *ww dd*. The percent of progeny with *W- D-* phenotype will be

   a.   0. •
   b.   25.
   c.   50.
   d.   75.
   e.   100.

4.   In *Drosophila*, the two genes *w* and *sn* are X-linked and 25 map units apart. A female fly of genotype $w^+ sn^+/w\ sn$ is crossed to a male from a wild type line. What percent of male progeny will be $w^+ sn$?

   a.   0
   b.   12.5 •
   c.   25
   d.   37.5
   e.   50

5.   In *Drosophila*, the genes for roughoid eyes (*r*) and javelin bristles (*j*) are linked 20 map units apart on one of the autosomes. If a dihybrid male fly $r^+ j/r\ j^+$ is test-crossed to an *rr jj* female, what proportion of progeny will be $r^+r\ j^+j$?

   a.   0 •
   b.   0.1
   c.   0.2
   d.   0.4
   e.   0.3

(Note to instructor: no crossing over in *Drosophila* males.)

6.   The mouse autosomal genes *B* and *S* are linked 38 map units apart. Genotypes *BB SS* and *bb ss* are intercrossed and the F_1 is testcrossed to *bb ss*. The proportion of *B- S-* progeny will be

   a.   0.38
   b.   0.76
   c.   0.5
   d.   0.31 •
   e.   0.19

7.   The maize genes *sh* and *bz* are linked 40 m.u. apart. If a plant $sh^+ bz/sh\ bz^+$ is selfed, what proportion of progeny will be *sh bz /sh bz*?

   a.   0.40
   b.   0.20
   c.   0.50
   d.   0.30
   e.   0.04 •

8. A new recessive mutation *d* in the plant *Arabidopsis* gives short stems. A mutant plant *dd* is crossed to another mutant *ee* (the *e* locus is known to be on chromosome 4). The F_1 was selfed and the F_2 was

0.56 *D- E-*
0.21 *D- ee*
0.21 *dd E-*
0.04 *dd ee*

The two loci are
   a.   unlinked.
   b.   linked 4 m.u. apart.
   c.   linked 20 m.u. apart.
   d.   linked 40 m.u. apart. •
   e.   linked 21 m.u. apart.

9. *A* and *B* are linked 20 m.u. apart, but *B* is on another chromosome. A plant of genotype *AA BB CC* is crossed to *aa bb cc* and the F_1 is testcrossed. What percentage of the progeny will be *Aa bb Cc*?
   a.   5 •
   b.   10
   c.   20
   d.   40
   e.   50

10. The *F*, *G*, and *H* loci are linked in the order written and there are 30 m.u. between *F* and *G* , and also 30 m.u. between *G* and *H*. If a plant *F G H/f g h* is testcrossed, what proportion of progeny plants will be *ff gg hh* if there is no interference?
   a.   0.70
   b.   0.30
   c.   0.245 •
   d.   0.15
   e.   0.21

11. Out of 800 progeny of a three-point test cross there were 16 double recombinants whereas 80 had been expected on the basis of no interference. The interference must have been
   a.   10%.
   b.   20%.
   c.   5%.
   d.   50%.
   e.   80%. •

12. In a four-point test cross the number of phenotypic classes of progeny is
    a.   16 •
    b.   8
    c.   81
    d.   4
    e.   depends on gene linkage

13. In a cross of two yeast strains of genotypes $c^+ a^+$ and $c\ a$, the progeny were

    40 $c^+ a^+$

    36 $c\ a$

    11 $c^+ a$

    13 $c\ a^+$

The recombinant frequency is
    a.   76%
    b.   12%
    c.   11%
    d.   13%
    e.   24% •

14. The following interlocus map distances were measured

    C to E 8 m.u.

    C to F 7 m.u.

    D to E 10 m.u.

The order of the genes must be
    a.   CDEF.
    b.   CFED.
    c.   DECF.
    d.   FECD.•
    e.   EDFC.

15. Consider a dog of the following genotype

    A   B   C
    ────────────────────────────────────
            30 m.u.   20 m.u.
    ────────────────────────────────────
    a   b   c

If this animal is testcrossed, what proportion of progeny will be *Aa Bb cc*?
    a.   0.07 •
    b.   0.30
    c.   0.02
    d.   0.15
    e.   0.10

Open-Ended Questions

1.   In rabbits the autosomal allele *B* gives black coat, and *b* gives brown. At another autosomal locus the allele *E* modifies the expression of *B* to give beige, and *E* also modifies *b* to give lemon. The allele *e* allows normal expression of black and brown. A beige female was crossed with a brown tester and the following progeny were obtained

black	183
lemon	181
beige	19
brown	17
Total	400

a.   What is the genotype of the beige female parent?
b.   What is the genotype of the brown tester male?
c.   Explain the numbers of the four phenotypes in the progeny.
d.   What must have been the genotypes of the two flies that were the parents of the beige female crossed?

Solution

a.   *Bb Ee*
b.   *bb ee*
c.   the two loci are linked 9 m.u. apart

$$\frac{B \quad\quad e}{b \quad\quad E}$$

and the brown and beige progeny must have been produced by crossovers (17 + 19 = 36 out of 400 = 9%)
d.   *BBee* and *bb EE*

2.   A fruitfly of genotype *Aa Bb* (parent 1) is crossed to another fruitfly of genotype *aa bb* (parent 2). The progeny of this cross were

Genotype	Number of individuals
Aa Bb	38
aa bb	37
Aa bb	12
aa Bb	13

a.   What gametes were produced by parent 1 and in what proportions?
b.   Do these proportions demonstrate independent assortment of the two genes?
c.   What can be deduced from these proportions?
d.   Draw chromosomal diagram(s) to illustrate the arrangement of the genes on the chromosomes in parents 1 and 2.
e.   Draw a diagram(s) to explain the origin of the two rarest progeny genotypes.

Solution

    a.   The proportions were 38% *AB*, 37% *ab*, 12% *Ab*, and 13% *aB*.

    b.   No: expect 25% of each under independent assortment.

    c.   The two gene pairs must be linked; RF = 12 + 13 = 25%, that is 25µ apart on the same chromosome pair.

    d.

A	*B*		*a*	*b*
a	*b*		*a*	*b*

    e.

A	*B*
A	*B*

         ×   <—— crossover at meiosis in parent 1.

a	*b*
a	*b*

3.   In the fruitfly *Drosophila*, the *A/a* and the *B/b* allele pairs are found on the same chromosome pair 20 centimorgans apart. A cross was made as follows:

A	*B*		*a*	*b*
a	*b*	×	*a*	*b*

What progeny genotypes are expected and in what proportions?

Solution

Aa Bb:	0.4
aa bb:	0.4
Aa bb:	0.1
aa Bb:	0.1

4.   The father of Mr. Spock, first officer of the starship Enterprise, came from the planet Vulcan, and Spock's mother came from Earth. Vulcans have pointed ears (allele *P*), and have no adrenal glands (allele *A*; — the basis of their coolness under pressure since they can produce no adrenalin). The corresponding Earth alleles are *p* and *a* respectively. The two allele pairs *P/p* and *A/a* are located on the same chromosome pair 16 centimorgans apart. If Mr. Spock mated with an Earth woman what is the probability that their first child would have both Vulcanian phenotypes (pointed ears and absent adrenal glands)?

Solution

The cross is *P A/p a* × *p a/p a*. Because total RF = 16%, the frequency of the *P A* gamete will be (100 - 16)/2 = 84/2 = 42%.

5.   In humans the allele *N* causes an abnormal shape of the patella in the knee (*n* is the normal allele). A separate gene is concerned with finger length, and the allele *B* causes abnormally short fingers whereas *b* gives normal length. A study focussed on people who had both abnormal patella and short fingers (they were most likely *Nn Bb*), having inherited the *N* allele from one parent and the *B* allele from the other parent. These *Nn Bb* people married normal spouses and 40 progeny from these matings were classified as follows:

normal	3
abnormal knees and fingers	2
abnormal knees only	17
abnormal fingers only	18

a.   Draw the chromosomes of the *Nn Bb* individuals, their parents, and their four types of children, showing the positions of the alleles.

b.   Explain why the four types of children were in the proportions shown.

Solution

a.   The results clearly depart from independent assortment, so the loci must be linked. The cross must have been

$$N\,b/n\,B \times n\,b/n\,b$$

and gametes

n b	3
N B	2
N b	17
n B	18

b.   Total RF = (3 + 2)/40 = 5/40 = 12.5% or 12.5 m.u.

6.   In a diploid species of rice a new recessive mutation arose that gave brown color to the interior of the seed (in wild types the color is white). Another mutation to brown seed interior had been obtained previously, and that one mapped to chromosome 2, at a distance of 40 map units from the locus for plant stature (d = dwarf, d^+ = wild type = tall). To determine if the new mutation was at the same or a new locus, a cross was made of the homozygous new brown mutation to a homozygous dwarf strain dd. The F_1 was wild type and this was selfed to obtain 100 F_2 individuals of the following types:

wild type	52
brown dwarf	4
brown	23
dwarf	21

a.   Draw chromosomes of the parents and F_1 showing positions of alleles assuming this is an allele of the previously mapped brown gene.

b.   Do these results tell us whether or not the new mutation is at the same locus as the previous brown mutation?

c.   Explain the phenotypic frequencies observed in the F_2.

Solution

a. $b\,d^+/b\,d^+ \times b^+\,d/b^+\,d \longrightarrow b\,d^+/b^+\,d$

b.   The frequency of *b d* gametes in the F_1 plant is expected to be $40/2 = 20\%$, so the frequency of *bb dd* (brown dwarf) offspring will be $0.2 \times 0.2 = 0.04$, precisely the frequency observed. If the new mutation were at a different locus it would be most likely not linked to *d*, and would show independent assortment giving a standard Mendelian dihybrid self expectation of 1/16 for the double recessive, which is 6.25%. It is also clear that the F_2 ratio is quite different from a 9:3:3:1.

7.   In maize, a strain homozygous for two recessive mutations, liguleless (*lg*) and glossy seedling (*gl*) was crossed to another strain homozygous for a dominant allele (*B*) that puts red pigment into the stem and leaves. The F_1 plant was backcrossed to the liguleless glossy parental strain, and seeds from that cross were of the following phenotypes and numbers:

lg	*gl*	+	172
+	+	*B*	162
lg	*gl*	*B*	56
+	+	+	48
lg	+	*B*	51
+	*gl*	+	43
lg	+	+	6
+	*gl*	*B*	5

a.   Make a linkage map of these three genes, showing map distances.
b.   Calculate any interference.

Solution

a.   Inspection of the classes shows clearly that all three loci are linked. The last two classes must be double recombinant classes and these show that the *gl* alleles have 'flipped' in relation to the parental chromosome arrangement, so *gl* must be in the middle and the gene order is as written. The *Rf* of *lg-gl* = $51 + 43 + 6 + 5/543$ = $105/543 = 19.3$ m.u.; the Rf of gl-B = $56 + 48 + 6 + 5/543 = 21.2\ \mu$. Overall:

lg	19.3	*gl*	21.3	*B*

b.   Expected double recombinants = $0.193 \times 0.213 \times 543 = 22.32$. Therefore the interference = $1 - (11/22.32)$ = approximately 50%.

8.   The recessive alleles *k* (kidney shaped eyes instead of round wild type), *c* (cardinal-colored eyes instead of wild-type red) and *e* (ebony body instead of the wild-type grey) identify three genes on chromosome 3 of *Drosophila*. Females with kidney-shaped, cardinal-colored eyes were mated with ebony males. The F_1 was wild type.

When F_1 females were test-crossed to *kk cc ee* males, the following progeny phenotypes were obtained

k	c	e	3
k	c	+	876
k	+	e	67
k	+	+	49
+	c	e	44
+	c	+	58
+	+	e	899
+	+	+	4
Total			2000

a. Determine the order of the genes and the map distances between them.
b. Draw the chromosomes of the parents and the F_1.
c. Calculate interference and indicate what you think of its significance.

Solution

a. There is a clear departure from independence. The double recombinants are the rarest classes, and inspecting these shows us that the allele pair that has 'flipped' compared with the parental gametes as that of the *e* locus, therefore the gene order must be *k-e-c*. The RF for *k-e* = 67 + 58 + 3 + 4/2000 = 6.6 μ. the *RF* for *e-c* = 49 + 44 + 3 + 4/2000 = 5 m.u. Overall

<u>*k*   6.6   *e*   5.0   *c*</u>

b. $k + c/k + c \times + e +/+ e+ \longrightarrow k + c/+ e +$

c. Expected number of double recombinants = 0.066 × 0.05 × 2000 = 6.6. This is very close to the observed value of 7, so there is no evidence for interference.

9. In the plant *Arabidopsis thaliana* the loci for pod length (*L* = long, *l* = short) and leaf hairs (*H* = hairy, *h* = smooth) are linked 16 map units apart on the same chromosome. The following crosses were made

(i) *LL HH* × *ll hh* ——> F_1
(ii) *LL hh* × *ll HH* ——> F_1

If the F_1s from (i) and (ii) are crossed
a. What proportion of the progeny is expected to be *ll hh*?
b. What proportion of the progeny is expected to be *Ll hh*?

Solution

a. The cross is *LH/lh* × *Lh/lH*. From the first parent the proportion of *lh* gametes is 1 - 0.84/2 = 0.42; from the other parent it is 0.16/2 = 0.08. So the frequency of *ll hh* progeny is 0.42 × 0.08 = 0.0336

b.   Two possibilities:
lh from left + *Lh* from right; $0.42 \times 0.42 = 0.1764$
Lh from left + *lh* from right; $0.08 \times 0.08 = 0.0064$

Total   0.1828 or 18.28%.

10.   Consider the following yeast linkage map (*ade*, *leu*, and *met* are all auxotrophic alleles).

ade-7   leu-2   met-3
_____
     5 m.u.   10 m.u.

If *ade-7 met-3* is crossed to *leu-2*, what proportion of progeny will be fully wild type if there is no interference?

Solution
The cross is   *ade*   +   *met*
                         ×
              +   *leu*   +

so half the double recombinants will be + + +. Their frequency will be $0.05 \times 0.10 \times 0.5 = 0.0025$.

11.   In the nematode *Caenorhabditis* the genes for eye color (alleles *B* black, *b* white) and body length (*L* long, *l* short) are linked 20 map units apart. A cross is made *B L* / *b l* × *B l* / *b L*. What proportion of the progeny will be white eyed and have short body?

Solution
Must have *bl* gamete from each parent.
$[(1 - 0.2)/2] \times [0.2/2] = 0.4 \times 0.1 = 0.04.$

12.   A female mouse from a pure wild-type line was crossed with a male mouse from a pure line having apricot eyes (*ap*) and grey coat (*gy*). The F_1 was all wild type in phenotype, and these were intercrossed to produce an F_2 that had the following composition

Females   all wild type
Males   45% wild type
          45% apricot grey
          5% grey
          5% apricot

a.   Explain these frequencies.
b.   Give the genotypes of the parents, the F_1 and F_2 males, and females.

Solution

    a.   Because of the sex difference there is clearly X-linkage of both genes.

    b. $++/++ \times a\,g \longrightarrow ++/a\,g$ and $++$. Crossing these gives us

Females   all $+\_+\_$
Males   0.45 $++$
            0.45 $a\,g$
            0.05 $+\,g$
            0.05 $a\,+$

The latter two classes arise from crossing over so $RF = 0.05 \times 2 = 0.10$.

13.   In the following cross the genes are inherited independently except for C and D which are tightly linked and show zero recombination.

     AA bb cc DD ee FF $\times$ aa BB CC dd EE ff

In the F_2 what proportion of individuals will be

    a.   pure breeding?
    b.   heterozygous at all loci?
    c.   homozygous recessive?
    d.   *Aa BB cc DD Ee ff*?
    e.   phenotype like either parent?
    f.   phenotype unlike either parent?
    g.   genotype like either parent?
    h.   genotype unlike either parent?

Solution

The F_1 is of the following constitution

       $A \ b \ \underline{c\,D} \ e \ F$

     $a \ B \ \overline{C\,d} \ E \ f$

We see that the C and D loci behave like one locus

    a. $1/2 \times 1/2 \times 1/2 \times 1/2 \times 1/2 = 1/32$.

    b. $1/2 \times 1/2 \times 1/2 \times 1/2 \times 1/2 = 1/32$.

    c.   zero because of the trans arrangement of c and d.

    d. $1/2 \times 1/4 \times 1/4 \times 1/2 \times 1/4 = 1/256$.

    e. $(3/4 \times 1/4 \times 1/4 \times 1/4 \times 3/4) + (1/4 \times 3/4 \times 1/4 \times 3/4 \times 1/4) = 18/1024$.

    f.   subtract answer to e from 1, answer $= 1006/1024$.

    g. $(1/4 \times 1/4 \times 1/4 \times 1/4 \times 1/4) + (1/4 \times 1/4 \times 1/4 \times 1/4 \times 1/4) = 2/1024 = 1/512$.

    h.   subtract answer to g from 1, answer $= 511/512$.

14.   In beans, tall (*T*) is dominant to short (*t*), red flowers (*R*) is dominant to white (*r*) and wide leaves (*W*) is dominant to narrow (*w*). The following cross is made and progeny obtained as shown

Cross   tall red wide × short white narrow

Progeny   478 tall red wide
                          21 tall white wide
                          19 short red wide
                         482 short white wide

Explain why only these progeny were obtained and in the proportions observed.

Solution
All progeny are wide so wide parent must be *WW*.
Short and white appear in progeny so parent was heterozygous for these genes.
Four phenotypes deviate from 1:1:1:1 ratio therefore genes linked.
The F_1 must have been $\underline{T \quad R}$ *W*   (may or may not be linked)
                                                $\underline{t \quad r}$ *W*
RF = 40/1000 = 4 m.u. between *T* and *R*.

6

Linkage II: Special Eukaryotic Chromosome Mapping Techniques

Multiple-Choice Questions

1. Ascospores from a cross *leu-2* × *ad-7 met-3* (all auxotrophic markers) are plated on minimal medium (no adenine, leucine, or methionine). If the three genes are unlinked, what proportion of ascospores is expected to grow into a colony on this medium?
    a. 0.2500
    b. 0.0050
    c. 0.0450
    d. 0.1250 •
    e. 0.5000

2. In one short chromosome arm the average frequency of crossovers per meiosis was meaured to be 0.5. What proportion of meioses will have no crossovers at all in this region?
    a. 0.90
    b. 0.74
    c. 0.60 •
    d. 0.52
    e. 0.40

3.   In a fungal cross $A B \times a b$ (the loci are linked), in meiosis in which there has been a four-strand double crossover between the loci, the percentage of meiotic products that will be recombinant for A and B is

    a.   0
    b.   25
    c.   50
    d.   75
    e.   100 •

4.   In a cross in which there is a mean crossover frequency of 0.9 between two marker loci, the recombinant frequency for those loci will be

    a.   90%
    b.   45%
    c.   30% •
    d.   10%
    e.   5%

5.   In a haploid organism the loci *leu* and *arg* are known to be linked 30 map units apart. In a cross $leu^+ arg \times leu arg^+$, what proportion of progeny will be *leu arg*?

    a.   0
    b.   0.05
    c.   0.10
    d.   0.15 •
    e.   0.30

6.   In a linear tetrad analysis, the second division segregation frequency of the *cyh* locus is 16%. The map distance from this locus to its centromere is

    a.   4 m.u.
    b.   8 m.u. •
    c.   16 m.u.
    d.   32 m.u.
    e.   50 m.u.

7.   In the cross $+ \times b$ the frequency of linear asci of the type

    (ascus stalk end) $+ + b b + + b b$

is 8%. The <u>first</u> division segregation frequency can be predicted to be

    a.   2%
    b.   4%
    c.   8%
    d.   16%
    e.   68% •

8.   A second-division segregation ascus is an indication of
      a.   a two-strand double crossover between centromere and locus.
      b.   a crossover between centromere and locus. •
      c.   meiotic nondisjunction in a fungus.
      d.   mitotic crossover.
      e.   interference.

9.   The maximum second division segregation frequency normally possible is
      a.   25%
      b.   33.3%
      c.   50%
      d.   66.7% •
      e.   100%

10.   In an unordered ascus analysis of the *Neurospora* loci *cot* and *fl*, there were 42 tetratypes and 8 nonparental ditypes out of a total of 290 asci. These loci are linked at a distance of
      a.   2 m.u.
      b.   5 m.u.
      c.   10 m.u. •
      d.   21 m.u.
      e.   25 m.u.

11.   In unordered tetrad analyses of two linked loci, a nonparental ditype is an indication of
      a.   no crossovers.
      b.   single crossover.
      c.   two chromatid double crossover.
      d.   three chromatid double crossover.
      e.   four chromatid double crossover. •

12.   In the cross $mt^+ al^+ \times mt\ al$, the *mt* locus shows an M_{II} frequency of 12%, the *al* locus an M_{II} frequncy of 10%. Most asci were parental ditypes. The order of loci is
      a.   *mt*-centromere-*al*.
      b.   *al*-*mt*-centromere.
      c.   centromere-*mt*-*al*.
      d.   centromere-*al*-*mt*. •
      e.   *al*-centromere-*mt*.

13. In *Drospohila*, the genes *B/b* (*B* = brown, *b* = black body) and *L/l* (*L* = long, *l* = short bristles) are linked in the order centromere-*L/l*-*B/b*. In a fly of genotype *B l/b L* a mitotic crossover occurs between the *B/b* and *L/l* loci. This could produce
    a.   single black spot on brown long background. •
    b.   single brown spot on black long background.
    c.   twin spots of black and short on brown long background.
    d.   twin spots of brown and long on black short background.
    e.   no spots on a brown long background.

14. In *Aspergillus* the following loci are linked in the order shown:
centromere-*pro-paba-bio-ylo*
In a diploid of genotype,
$pro^+ \ paba^+ \ bio^+ \ ylo^+ / pro \ paba \ bio \ ylo$,
all wild-type alleles are dominant. A yellow diploid sector was found and isolated and upon test was found to be *paba* and *bio* requiring. A mitotic crossover must have occurred
    a.   distal to *ylo*.
    b.   between centromere and *pro*.
    c.   distal to *pro*.
    d.   between *pro* and *paba*. •
    e.   between *bio* and *ylo*.

15. From an *Aspergillus* diploid that was *w a b c d/+ + + +*, white (*w*) haploidized sectors were isolated. There were equal numbers of the following four genotypes:
abcd, abc+, +bcd, +bc+
From these data it can be concluded that the loci on the same chromosome as *w* are
    a.   *b* and *c*. •
    b.   *b* and *c* and *d*.
    c.   *a* and *d*.
    d.   *a* and *c*.
    e.   *a* and *c* and *d*.

Open-Ended Questions

1. The linkage of two *Neurospora* genes *cyh* and *al-2* is shown in the following diagram. Also shown are the regions of the chromosome which have been given Roman numbers. CEN = centromere.

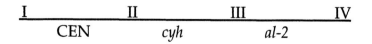

Draw linear octads to represent meioses in in a cross + + × *cyh al-2* in which there has been

    a.   no crossovers at all in the regions designated.
    b.   a single crossover in I.
    c.   a single crossover in II.
    d.   a single crossover in III.
    e.   a single crossover in IV.
    f.   a single crossover in each of II and III both involving the same pair of chromatids.

Solution
(Note: only tetrads shown—these should be doubled to obtain octads.)
    a.   *cyh al / cyh al / + + / + +*
    b.   *cyh al / cyh al / + + / + +*
    c.   *cyh al / + + / cyh al / + +*
    d.   *cyh al / cyh + / + al / + +*
    e.   *cyh al / cyh al / + + / + +*
    f.   *cyh al / + al / cyh + / + +*
(Note: half asci can have inverted order.)

2.   An *Aspergillus* diploid is constructed from a wild-type strain and one of genotype *ad his met ylo*. This diploid is wild-type in phenotype, but yellow (*ylo*) diploid sectors appear on it . These sectors were isolated and tested for their ability to grow on various media and the results were as follows where M means methionine, H means histidine and A means adenine.

	Min	M + H	M + A	H + A	M + H + A
Type 1	+	+	+	+	+
2	-	-	+	+	+
3	-	-	-	+	+
4	-	-	-	-	+

There were appreciable numbers of each type. Explain what these four types are and show how each was produced.

Solution

Type 1	+	+	+
2	*ad*	+	+
3	*ad*	*his*	+
4	*ad*	*his*	*met*

This shows that homozygosity for *ylo* can be accompanied by homozygosity for *ad*, for *ad* and *his*, or for *ad* and *his* and *met*. This shows the gene order is *met-his-ad-ylo*.

3.   In *Aspergillus*, the genes *y* (yellow), *os* (osmotic), *c* (colonial), *leu* (leucine), and *a* (aconidial) are known to be on the long arm of chromosome 3. A diploid was made between a haploid wild-type and a haploid of genotype *y os c leu a*. From the green diploid, yellow diploid sectors appeared, and these were sampled and tested with the following results

56% yellow osmotic aconidial leucine
40% yellow osmotic colonial aconidial leucine
4% yellow aconidial leucine

    a.   What is the origin of the three types of yellow sectors? (Diagram it.)
    b.   What can you tell about the order and position of the five loci?

Solution
    a.   Mitotic crossovers at different positions along the arm is the origin.
    b.   The gene order and relative distance must be as follows (*y*, *a*, and *leu* must be close together).

    (*y, a, leu*)-4-*os*-56-*c*-40-centromere

4.   In the fungus *Neurospora*, a cross was made between an auxotrophic strain mutant at the *arg-6* gene and an albino strain mutant at the *al2* locus. An analysis of 100 linear tetrads was performed and the results are below. (Each genotype represents an ascospore pair.)

arg-6	+	+	al2	arg-6	+	arg-6	+	+	al2
arg-6	+	+	al2	+	al2	+	al2	arg-6	+
+	al2	arg-6	+	arg-6	+	+	al2	arg-6	+
+	al2	arg-6	+	+	al2	arg-6	+	+	al2
38		**37**		**6**		**6**		**5**	

+	al2	arg-6	+
arg-6	+	arg-6	al2
+	al2	+	+
arg-6	+	+	al2
7		**1**	

    a.   Draw a map of these loci and their centromere(s). Show map distances.
    b.   Why are the first two types of tetrads equal in frequency?
    c.   Draw a diagram of the meiotic events that gave rise to the last type, and state why it was so rare.

Solution
    a.   centromere——12 m.u.*arg-6*-0.5-*al2*
    b.   Because of random spindle attachment at the first anaphase of meiosis.
    c.   All asci are parental ditypes except this one which shows a crossover between the loci, resulting in a tetratype.

5.   In a fungal cross $a\ b\ \times + +$, gene a is on chromosome 1 and gene b is on chromosome 6. It is known that the two loci are so closely linked to their centromeres that there is never any crossing over in either centromere to locus region. What proportion of linear tetrads will be

    a. $M_I\, M_I$ PD

    b. $M_I\, M_I$ NPD.

    c. $M_I\, M_{II}$ T

    d. $M_{II}\, M_I$ T

    e. $M_{II}\, M_{II}$ PD

    f. $M_{II}\, M_{II}$ NPD

    g. $M_{II}\, M_{II}$ T

Solution

Because there is no crossing over in either gene-to-centromere region there can be no M_{II} asci.

    a.   0.5

    b.   0.5

    c.   0

    d.   0

    e.   0

    f.   0

6.   *Penicillium* is a common bluish mold from which the antibiotic penicillin is extracted. A *Penicillium* diploid was made by fusing haploid strains of genotype

$ad^+\ leu^+\ ribo^+\ fpa^+\ w^+$ and *ad leu ribo fpa w*.

The mutant alleles are all recessive and determine requirements for adenine, leucine, and riboflavine; resistance to fluorophentlalanine; and white asexual spores, respectively. When thousands of asexual spores of the diplod straind were spread on medium containing fluorophenylalanine, adenine, leucine, and riboflavine, only 93 colonies grew and these were isolated.

    a.   It was determined that 38 colonies were haploid; 22 with the genotype *fpa leu ribo ad$^+$ w$^+$*, and 16 with the genotype *fpa leu ribo ad w*. What does this result tell us about the linkage of the genes? Summarize with a diagram.

    b.   The remaining 55 colonies were diploid; of these 24 required neither leucine nor riboflavine, 17 required riboflavine but not leucine, and 14 required both. What further linkage information does this result provide? Summarize with a diagram showing positions and relative distances.

Solution

    a.   Haploidization to produce *fpa* cells always results in *leu ribo* cells too, so these must be linked to *fpa*. The markers *ad* and *w* are also linked and together assort independently of the other chromosome.

(*fpa/leu ribo*)   (*ad/w*)
order unknown

b.   The data show homozygosity through mitotic crossovers can be for *fpa* alone, or *fpa* and *ribo*, or for *fpa* and *ribo* and *leu*, therefore order and distance is
fpa   24   *ribo*   17   *leu*   14   centromere

7.   In the fungus *Neurospora*, a strain that was auxotrophic for thiamine (mutant allele *t*) was crossed to a strain that was auxotrophic for methionine (mutant allele *m*). Linear asci were isolated and these were classified into the following groups.

Spore pair			Ascus types			
1 and 2	*t* +	*t* +	*t* +	*t* +	*t m*	*t m*
3 and 4	*t* +	*t m*	+ *m*	+ +	*t m*	+ +
5 and 6	+ *m*	+ +	*t* +	*t m*	+ +	*t* +
7 and 8	+ *m*	+ *m*	+ *m*	+ *m*	+ +	+ *m*
Number	260	76	4	54	1	5

a.   Determine the linkage relationships of these two genes to their centromere(s) and to each other. Specify distances in map units.
b.   Draw diagrams to show the origin of each ascus type.
c.   If 1000 randomly selected ascospores are plated on a plate of minimal medium, how many are expected to grow into colonies?

Solution
a. $M_{II}(t) = (4 + 54 + 5/2)/400 = 8$ m.u. $M_{II}(m) = (85/2)/400 = 12.1$ m.u.
The RF = T/2 + NPD = [(76 + 54 + 5)/2 + 1]/400 = 17.1 m.u.
Hence loci are linked on oposite sides of centromere. (Nonadditivity results from the double crossover asci.)
t-8-cen-12.1-*m*
b.   In order from left: - no crossover; SCO (*m* side); 2 strand DCO; SCO (*t* side); 4 strand DCO; 3 strand DCO.
c   RF = 17.1%, and only 1/2 these will be + + prototrophs, that is, 8.6% or 86 out of 1000.

8.   In *Drosophila*, the genes for ebony body (*e*) and stubby bristles (*s*) are linked on chromosome 2. A fly of genotype + *s/e* + developed predominantly wild-type but showed two interesting aberrations. The first was a pair of adjacent patches, one with stubby bristles and the other with ebony body. The second aberration was a solitary patch of ebony color. What are the likely origins of these two types of unexpected aberrations?

Solution
The genes must be in the order *e-s*-centromere. A mitotic crossover between the centromere and *s* would give the twin spot. A mitotic crossover between *s* and *e* would give the single ebony spot.

9.   In a haploid fungus a random ascospore analysis of the cross

   leu arg$^+$ $\times$ *leu*$^+$ *arg*

gave a recombinant fraction of 0.2 (20%). For a certain experimental purpose the researchers wanted to obtain tetratype asci from this cross.

   a   Draw a tetratype ascus from this cross.
   b   Given the RF obtained, what will be the frequency of tetratype asci expected from this cross? (Make the simplifying assumption that the only meiotic classes of any appreciable frequency will be those with 0, 1, or 2 crossovers between the two loci.)

Solution

   a.   *leu arg* $^+$/*leu arg*/*leu*$^+$*arg*$^+$/*leu*$^+$*arg*
   b.   Using the mapping function we find that e^{-m} = 0.6, so m = 0.513. Now we calculate the sizes of the 0, 1, and 2 crossover classes using the Poisson formula. In summary

Crossover	Frequency	Tetratypes
0	0.6	0
1	0.31	0.31
2	0.08	0.04
	Total	0.35

   Hence 35% of asci will be tetratype.

10.   What fungal genotypes would force the formation of a trikaryon, a fungal strain that has three different nuclear types in a common cytoplasm?

Solution
Three haploid strains of the following types:

aux1	*aux2*	+
aux1	+`	*aux3*
+	*aux2*	*aux3*

C H A P T E R

7

Gene Mutation

Multiple-Choice Questions

1. A mutation that occurred in a plant petal would be best termed
    a. germinal.
    b. somatic. •
    c. suppressor.
    d. dominant.
    e. recessive.

2. Forty units of enzyme A are needed to produce wild-type phenotype. The wild-type allele produces 20 units and a new mutation produces only five units. The mutation would be
    a. dominant. •
    b. recessive.
    c. incompletely dominant.
    d. codominant.
    e. overdominant.

3. An individual cell homozygous for a certain mutant allele of the gene for the human enzyme phenylalanine hydroxylase contains no detectable activity for that enzyme, The mutation is best described as
    a. dominant.
    b. recessive.
    c. prototrophic.
    d. gain of function.
    e. null. •

4.   Fifty cells of an auxotrophic strain of haploid yeast are plated on minimal medium. The number of colonies expected to grow is
   a.   0. •
   b.   1.
   c.   25.
   d.   50.
   e.   unknown.

5.   A new *Neurospora* mutant was tested for auxotrophy. (Abbreviations: minimal medium = M, arginine = A, proline = P, histidine = H). The mutant grew on M + A + H, M + P + H, and M + A + H + P, but not on M or M + A + P. The mutant requires
   a.   arginine.
   b.   histidine. •
   c.   proline.
   d.   histidine and proline.
   e.   arginine and proline.

6.   The *arg-1a* allele of *Aspergillus* reverts at a frequency of 10^{-4}. If a million haploid *arg-1a* cells are plated on minimal medium, the number of colonies expected is
   a. 0.
   b. 4.
   c. 10.
   d. 100. •
   e. 1000.

7.   Two new mutations auxotrophic for cysteine occur in different haploid yeast strains. If the mutations are in two different genes on seperate chromosomes, when the strains are crossed the proportion of prototrophic progeny should be
   a.   0.
   b.   0.25. •
   c.   0.5.
   d.   0.75.
   e.   1.00.

8.   A bacterial histidine mutant was plated on minimal medium. A single colony grew. This must have arisen from
   a.   forward mutation.
   b.   auxotrophic mutation.
   c.   back mutation.
   d.   suppressor mutation.
   e.   back mutation or suppressor mutation. •

9. Twenty phage particles are mixed with a dense suspension of bacteria and the mixture is immediately spread on medium in a petri dish. The number of plaques expected is
    a. 0.
    b. 1.
    c. 20. •
    d. large numbers.
    e. unknown.

10. The ClB test in *Drosophila* detects
    a. crossover suppressor mutations.
    b. any mutations in the gene represented by 'l'.
    c. Barr-eye mutation.
    d. any lethal mutation on the X chromosome. •
    e. mutations in either C, l, or B.

11. People who develop the hereditary form of retinoblastoma inherit
    a. a single dominant allele causing retinoblastoma.
    b. a single recessive allele causing retinoblastoma. •
    c. a pair of dominant alleles causing retinoblastoma.
    d. a pair of recessive alleles causing retinoblastoma.
    e. a pair of normal (nonretinoblastoma) alleles.

12. A mutant allele of corn stands for absence of red anthocyanin pigment in the kernel (absence of anthocyanin results in a yellow color). However, the allele is unstable, reverting quite late, yet often, in development. The expected phenotype of kernels homozygous for this mutation is
    a. fully red.
    b. yellow with many large red spots.
    c. yellow with few large red spots.
    d. yellow with many small red spots. •
    e. yellow with many large red spots.

13. The sequence of mutations leading to nonhereditary retinoblastoma (retinal cancer) can best be written (where *r* or *R* could be the mutant allele)
    a. *rr->RR.*
    b. *Rr->rr.*
    c. *RR->Rr->rr.* •
    d. *RR->rr.*
    e. *rr->Rr->RR.*

14. A *Neurospora* adenine auxotroph became prototrophic by a suppressor mutation 25 map units away from the original mutation. If this strain is crossed to wild-type, the percent of adenine-requiring progeny will be
   a. 0.
   b. 12.5. •
   c. 25.
   d. 50.
   e. 75.

15. Brachydactyly (short fingers is caused by a dominant mutation with full penetrance. In one town over a number of years 25,000 children were born; and of these, 50 had brachydactyly. However, 48 of these children had a brachydactylous parent. The frequency of new mutations per gamete is
   a. 2×10^{-4}
   b. 2.5×10^{-4}
   c. 4×10^{-6}
   d. 8×10^{-5}
   e. 4×10^{-5} •

Open-Ended Questions

1. In a colony of mice, a mutant allele arose that made the coat a slate grey color and also produced several mild skeletal anomalies. When pure-breeding, greys were crossed to wild-type. The progeny were always wild-type and, when these heterozygotes were mated, the progeny were approximately 80% wild-type and 20% grey.
   a. Is the allele dominant or recessive?
   b. Explain the approximately 80:20 ratio.

Solution
   a. The heterozygotes are wild so grey is recessive; it is given the symbol *gl*.
   b. It is probable that the allele is sublethal. In other words the expected ratio of 75%:25% is modified to 80:20 because of inviability of the grey progeny. Out of the expected 25, 5 die, so there is a 25% inviability. The effect seems unlikely to be due to the grey color, but clearly the allele is pleiotropic, affecting several aspects of the mouse phenotype.

2. The mouse mutation *T* gives a short tail (brachyuria). Several *T*/+ heterozygotes were allowed to cross and sacrificed to examine embryos in utero. It was found that there were 53 normal-looking embryos, 97 with brachyuria, and 48 showed small size and extreme skeletal abnormalities. Propose a hypothesis to explain these results and state how you would test it.

Solution
TT individuals are probably lethal. If true, the live progeny should be 2/3 brachyuric and 1/3 wild-type phenotype.

3.   In fungal genetics, heterokaryons are sometimes used to help sickly mutations through meiosis. For example, many mutations of the *ba* gene are barren as females, but in *ba*+/*ba* heterokaryons the *ba*+ gene product passes through the common cytoplasm and allows some *ba* nuclei to participate in meiosis.

A heterokaryon   (*A ba cyh^R ad-1 leu-2$^+$*)

plus

(*A ba$^+$ cyh^S ad-1$^+$ leu-2*)

is crossed to a wild-type strain of mating type *a*.
(*A* and *a* are mating type alleles, *cyh* is a gene for cycloheximide resistance, *ad* is for adenine requirement and *leu* is for leucine requirement.)
Devise a simple plating method to isolate *ba* ascospores that have come through meiosis. Note that meioses involving *ba* might be very rare.

Solution

The *ba* nucleus carries a *cyh^R* allele so if the ascospores are plated on cycloheximide, any that grow must have come from a meiosis involving the *ba* nucleus. A proportion of these ascospores (probably <1/2) will be *ba* in genotype.

4.   A population of 100 female mice was exposed to high doses of gamma rays. What simple protocol can be used to investigate whether the radiation induced recessive lethal mutations on the X chromosomes?

Solution
Cross the females repeatedly to wild-type males and in the combined progeny measure the sex ratio. An excess of females would indicate the occurrence of X-linked recessive lethels in the gonads of the exposed females.

5.   In a ClB test a wild-type male is irradiated with X-rays and crossed to a ClB/X$^+$ female. One daughter with bar eyes is selected and crossed to a wild-type male. The progeny are

  1/3 ClB females
  1/3 wild-type females
  1/3 dwarf males

Explain this ratio and the genetic effect of the X-rays.

Solution
The X-rays must have produced an X-linked recessive allele *d* in one sperm, eventually causing the dwarf phenotype.

$X^+/Y \quad \times \quad ClB/X^+$

$ClB/X^d \times X^+/Y$

Progeny
- 1/4 ClB/X⁺ (ClB female)
- 1/4 ClB/Y (lethal)
- 1/4 X⁺Xᵈ (wild female)
- 1/4 XᵈY (dwarf male)

6. Testicular feminization syndrome (TFS) is inherited as an X-linked recessive phenotype. In a certain pedigree there was no record of anyone ever having TFS, but one couple's first two children had TFS, and their third was a normal girl. The girl is now married and is contemplating having a child.
   a. What is the reason the normal pedigree suddenly showed two cases of TFS? (Draw out the pedigree showing genotypes.)
   b. What advice would you give her about childbearing children?

Solution
   a. A mutation to the TFS allele must have occurred either in the woman's germinal tissue or in her parents' germinal tissue. Therefore a large proportion of her germinal cells must be X^+X^t; the husband must be X^+Y, and the first two sons must be $X^t Y$.
   b. The daughter has up to a 50% chance of being X^+X^t, in which case up to 1/2 her sons would have TFS, but no daughters.

7. In yeast, wild-type colonies are white, but adenine-requiring auxotrophs (*ad*) are red when grown on medium containing limiting amounts of adenine due to the accumulation of a red precursor in adenine biosynthesis. If cells from an *ad* strain are plated on minimal medium most cells don't produce colonies but two rare types of 'revertant' colonies are produced, one type white in color, and the other type with a 'rose' color on limiting adenine. Strains grown from such colonies were crossed to wild-type of opposite mating type and the results were as follows:

white 'revertant' × wild-type (also white)
Progeny: all white auxotrophs

rose 'revertant' × wild-type (white)
Progeny: 1/2 white
1/4 rose
1/4 red

a. What is the difference between the genetic mechanisms that produced the white and the rose colonies?

b. Show the genotypes of the parents and progeny in the above crosses.

Solution

a. The white colony looks and crosses like a geniune revertant. However, the rose colony throws off genuine adenine mutants when crossed to wild-type so the *ad* mutation must still have been there but in a partially suppressed state.

b. The white revertant is + (= *ad*+), the rose colony is *ad su* which on crossing to + + gives equal numbers of + + , *ad* +, + *su*, and *ad su* which are white, red, white, and rose, respectively.

8. To obtain revertants, millions of cells of a histidine-requiring (*his*) auxotrophic mutant haploid strain of the fungus *Ascobolus immersus* were plated on minimal medium. Two types of colonies were obtained, a large smooth type similar to that produced by wild-type cells, and also some that were ragged-looking with uneven edges. A ragged strain was removed from the plate and crossed to a wild-type of opposite mating type. Unordered asci were isolated and tested on minimal medium, and the results were as follows:

1	2	3
ragged	ragged	no growth
no growth	ragged	no growth
smooth	smooth	smooth
smooth	smooth	smooth
2 asci	11 asci	12 asci

a. Explain the origin of the ragged strains generally.

b. Explain the origin and frequencies of the three ascus types, showing parent and ascospore genotypes.

c. What do you think was the origin of the smooth revertants? How would you test your idea?

Solution

a. Most likely suppressed *his* mutant because the crosses regenerate his progeny (no growth on minimal medium).

b. The cross was *his sup* (ragged) × + + (smooth) which would give three types of unordered tetrads, tetratype (type 1), parental ditype (2), and nonparental ditype (3). Because PD + NPD the suppressor was probably not linked to the *his* gene.

c. Probably true revertants.

19. You wish to measure the reversion rate of a *gal4* mutation in haploid yeast. The *gal4* mutants are unable to grow when galactose is the only carbon source, so you can detect revertants by plating on galactose plates. You innoculate each of 50 small tubes with 1 mL of glucose-based medium and about 10 *gal4* cells, and incubate the tubes until they each contain about 2×10^7 cells. You then plate the contents of each tube on a single galactose plate, and count *gal4+* colonies after incubation.

Number of colonies on each plate:

1	5	0	2	3	1	4	1	0	2
2	0	0	3	0	2	123	0	1	0
0	1	1	0	0	2	1	0	1	0
2	1	0	1	0	1	0	2	3	0
0	27	1	3	0	1	0	1	0	2

Calculate the reversion rate of the *gal4* mutation.

Solution

We need the Poisson formula $f(0) = e^{-un}$ and solving

$$20/50 = 0.4 = e^{-u} \times 2 \times 10^7$$

$$\ln 0.4 = -u \times 2 \times 10^{-7}$$

$$u = 0.916 / 2 \times 10^7$$

we find that $u = 0.458 \times 10^{-7}$ or 4.58×10^{-8}.

10. A lab population of *Drosophila* was started with a small number of males and females from the standard pure line called *oregon R*. After several months the population had increased to hundreds of thousands of flies. Careful screening of individual flies under the microscope revealed many different mutant phenotypes. Three examples were
1. A male with forked bristles (wild-type is unforked).
2. A female with apricot eyes (wild-type is red).
3. A male with curled wings (wild-type is straight).

Crosses were made with the following results
1. forked ♀ × wild ♂

F_1 all wildtype

F_2 3/4 wild-type (males and females)

    1/4 forked (males and females)

2. apricot ♀ × wild♂

F_1 males all apricot
    females all wild

F_2 1/2 wild-type (males and females)
    1/2 apricot (males and females)

3. curled ♂ (male) × wild ♀

F_1 1/2 curled (males and females)
    1/2 wild-type (males and females)

    a.   Explain what the results tell us about the nature of these three mutations.
    b.   Speculate on whether the mutations could have occurred in the individuals chosen, or in their immediate or distant ancestors.

Solution
    a.   The forked phenotype behaves like a standard autosomal recessive.
           ff × ++ -> +*f* -> 3/4 +_ and 1/4 *ff*

The apricot phenotype is X-linked recessive
           aa × +Y -> +*a* females and *a*Y males -> 1:1 in both sexes in F_2

Curley acts like a dominant, and the individual crossed must have been heterozygous
           C+ × ++ -> 1/2 C+ and 1/2 ++

    b.   All the mutations must have been in the germ lines of previous generations. The recessives may have been brought to homozygosity by inbreeding.

C H A P T E R

8

Chromosome Mutation I: Changes in Chromosome Structure

Multiple-Choice Questions

1.   In an animal bearing the heterozygous inversion *ABCDE • FGHI/ABGF • EDCHI*, in one meiocyte a crossover occurred between the *D* and *E* loci and another crossover occurred between the *F* and *G* loci. The two crossovers involved the same two chromatids. What will be the proportion of abnormal meiotic products from that meiosis?
    a.   0% •
    b.   25%
    c.   50%
    d.   100%
    e.   depends on the size of the *D–E* and *F–G* regions

2.   In an animal bearing the heterozygous inversion *ABCDE • FGHI/ABGF • EDCHI*, in one meiocyte a crossover occurred between the *D* and *E* loci and another crossover occurred between the *F* and *G* loci. The two crossovers involved all four chromatids. What will be the proportion of abnormal meiotic products from that meiosis?
    a.   0%
    b.   25%
    c.   50%
    d.   100% •
    e.   depends on the size of the *D–E* and *F–G* regions

3.   A wild-type chromosome can be represented as *ABC • DEFGH*, and from this a chromosomal aberration arises that can be represented *ABC • DEGFH*. This is known as
    a.   deletion.
    b.   translocation.
    c.   duplication.
    d.   pericentric inversion.
    e.   paracentric inversion. •

4.   A pure line of plants of genotype *aa bb cc dd ee* (all recessive to wild-type) was crossed to a wild-type. One F_1 individual expressed the recessive alleles *d* and *e*. This individual arose from
    a.   gene mutation in wild-type parent.
    b.   reversion in the quadruple mutant.
    c.   deletion in the quadruple mutant.
    d.   deletion in the wild-type. •
    e.   position effect variegation.

5.   A female *Drosophila* supposedly heterozygous for two recessive mutations *cn* and *lz* that are on the same arm of the X chromosome (*cn lz*/+ +) surprisingly expresses both these genes. The male progeny of this female will be
    a.   all wild-type.
    b.   all *cn lz*. •
    c.   1/2 *cn lz* and 1/2 wild-type.
    d.   *cn* +.
    e.   + *lz*.

6.   A chromosome aberration of the type *ABCD • EFFGH* is called
    a.   inversion.
    b.   translocation.
    c.   deletion.
    d.   duplication. •
    e.   dicentric.

7.   A meiocyte of an organism heterozygous for a reciprocal translocation goes through meiosis and results in four viable meiotic products. This means
    a.   the translocation reverted.
    b.   a suppressor mutation occurred.
    c.   there was adjacent segregation.
    d.   there was alternate segregation. •
    e.   translocation breakpoints were very close to centromeres.

8.   Acentric fragments are produced in
   a.   paracentric inversions. •
   b.   pericentric inversions.
   c.   translocations.
   d.   deletions.
   e.   duplications.

9.   The two loci y/y and R/r are normally 20 m.u. apaprt. In a heterozygous inversion the two loci are just within the breakpoints of the inversion. The RF between these loci when dihybrid in such an inversion heterozygote will be
   a.   20%.
   b   slightly less than 20%.
   c.   slightly more than 20%.
   d.   10 %.
   e.   0. •

10.   A diagnostic of a reciprocal translocation is
   a.   two genes on different chromosomes show *RF* of 50%.
   b.   two genes on different chromosomes show *RF* of 0%.
   c.   two genes on different chromosomes show linkage. •
   d.   *RF* of linked genes drops to zero.
   e.   *RF* of linked genes decreases.

11.   A fungal cross between two linked loci *ad pan* × + + gave only the following two types of tetrads

Spore 1:	*ad pan*	*ad pan*
Spore 2:	aborted	*ad* +
Spore 3:	aborted	+ *pan*
Spore 4:	+ +	+ +
	type 1	type 2

These results suggest
   a.   a deletion in the wild-type parent.
   b.   a heterozygous translocation with adjacent segregation.
   c.   a heterozygous translocation with alternate segregation.
   d.   a heterozygous inversion. •
   e.   a duplication.

12.   A set of *Drosophila* deletions was assembled as follows

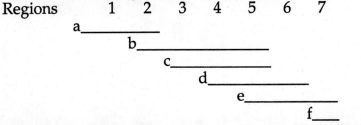

A radioactive DNA fragment containing a certain gene bound only to chromosomes a and f. The gene must be in region
- a.   1
- b.   2
- c.   3
- d.   4
- e.   5 •

13.   A *Neurospora* cross gave the following three types of octads only

back/black/black/black/black/black/black/black
abort/abort/abort/abort/abort/abort/abort/abort
black/black/abort/abort/black/black/abort/abort

This suggests a
- a.   heterozygous deletion.
- b.   heterozygous duplication.
- c.   heterozygous pericentric inversion.
- d.   homozygous inversion (either kind).
- e.   heyterozygous reciprocal translocation. •

14.   In yeast, the gene *phe1* is on chromosome 3 and *lys4* is on chomosome 7. A cross is made + + × *phe1 lys4* and the progeny were

47% + +
44 % *phe1 lys4*
4% *phe1* +
5% + *lys4*

This suggests
- a.   position effect variegation.
- b.   a tanslocation. •
- c.   a terminal deletion.
- d.   a pericentric inversion.
- e.   unequal crossovers.

15.   The *Drosophila* locus *rst* is normally located at the end of a chromosome arm. The recessive allele *rst* gives a rough appearance to the eyes. In a certain *rst*/+ heterozygote the + allele is relocated next to the centromere by a paracentric inversion. This fly showed patches of rough and smooth areas in its eyes. This effect is called
- a.   semisterility.
- b.   crossover suppression.
- c.   position effect variegation. •
- d.   mitotic crossing over.
- e.   centromeric fusion.

Open-Ended Questions

1.   The human genetic disease tuberous sclerosis (TS) is inherited as an autosomal dominant. Polycistic kidney disease (PKD) is also inherited as an autosomal dominant. Both diseases are dominant because of haplo-insufficiency; in other words in heterozygotes the one wild-type allele is insufficient for normal cell function. The genes concerned are closely linked on the short arm of chromosome 16. Rare cases have been reported of people expressing the symptoms of both TS and PKD. In these cases the two diseases are inherited together as one unit through the generations.

    a.   Propose a possible mechanism for the origin of the people who have both diseases, and explain how they are inherited together.

    b.   How would your idea be tested?

Solution

    a.   It could be double mutation, but more likely it is a microdeletion spanning both loci. Because of haplo-insufficiency the deletion would act as a dominant disease allele for both loci. The deletion would be inherited as a unit keeping the two diseases together in subsequent generations.

    b.   Check pedigrees: in new cases the two diseases should arise together. The deletion might be visible microscopically but probably not. The best test for the deletion would be molecular; use probes for both the TS and PKD genes and find no in situ hybridization to one homolog.

2.   In a lab strain of *Drosophila*, cinnabar (*cn*) and brown (*bw*) are recessive eye color mutations known to be 41 map units apart on chromosome 2. When similar mutant alleles were induced in a strain from nature, the same linkage of *cn* and *bw* was observed. However, when a wild-type strain from nature was crossed with a *cn bw / cn bw* lab strain to create the genotype + + / *cn bw*, and females of this type were test-crossed to *cn bw / cn bw* males from the lab strain, the following phenotypic proportions were obtained in the progeny

+	+	25,200
cn	*bw*	21,009
cn	+	11
+	*bw*	36

    a.   What is unexpected about these results?

    b.   What is the most likely explanation? (Summarize your model with a chromosome diagram with labelled genes.)

    c.   On your diagram show the precise origin of the *cn* + and the + *bw* classes.

Solution

    a.   The *RF* is much less than the expected 41%.

    b.   There must be a large inversion in the *cn* to *bw* region.

Strain from nature   *cn*—[inversion spanning most of region] — *bw*
Standard lab strain + —[noninverted sequence] — +
These two chromosomes are paired in the F_1, with the inverted region looped.

    c.   These genotypes must arise from crossovers in the small homologous noninverted regions shown with dashes in the diagram.

3.   In corn, the genes for tassel length (alleles *T* and *t*) and rust resistance (alleles *R* and *r*) are known to be on separate chromosomes. In the course of making routine crosses, a breeder noticed that one *Tt Rr* plant gave unusual results in a test cross to a the double recessive pollen parent *tt rr*. The results were

Progeny	*Tt Rr*	98
	tt rr	104
	Tt rr	3
	tt Rr	5

Corn cobs: only about half as many seeds as usual.

    a.   What are the key features of the data that are different from expected results?
    b.   State a concise hypothesis that explains the results.
(Include a diagram showing the arrangement of alleles on the chromosomes.)
    c.   Explain the origin of the two classes of progeny with 3 and 5 members.

Solution

    a.   One expects equal numbers of the four phenotypes but there is a deficiency of *Tr* and *tR*, suggesting a translocation. This inference is supported by the semisterility of the plant.
    b.   There are several arrangements that fit but here is one

Also *T* and *R* could be on the translocation pair and *t* and *r* on normal chromosomes. The main types of gametes that will survive from these possibilities are *TR* and *tr*.
    c.   These arise from crossovers in the region from *T/t* to the translocation point or from *R/r* to the translocation point.

4.   In the fungus *Neurospora* the *cyh* locus is 10 map units from the centromere on chromosome 1 and the *tryp* locus is 4 map units from the centromere on chromosome 7.

A double mutant strain *cyh tryp* was crossed to a *cyh*+ *tryp*+ strain isolated from a rotting coconut. Three hundred ascospores were examined and tested and they were of the the following types (the *cyh* and *tryp* alleles of the aborted ascospores obviously could not be tested).

cyh^+ $tryp^+$	74
cyh $tryp$	70
cyh^+ $tryp$	3
cyh $tryp^+$	7
white (aborted)	146

a.   Provide a general explanation of these results saying what (if anything) is unusual about them.

b.   Explain the origin of each of the five categories of ascospores using a diagram.

Solution

a.   About 50% abortion means semisterility so a translocation is suggested. This is supported by an *RF* of only 10/330 = 3.3% instead of the expected 50% from the independent assortment.

b.   The heterozygous cross can be represented

N1 _____ c _____   N4 - - - - - - - - - - - - - - - - t - - -
T1 _____ + ____ - - - - - - - -   T4 _____ - - + - - -

and this would pair as a cross at meiosis.

The majority of spores will be N1 N2 (*ct*) or T1 T4 (++). Some +*t* and *c*+ types will arise from crossing over in the locus-to-breakpoint regions.

5.   Given a reciprocal translocation in *Neurospora*: you know that you expect 50% of your progeny from a cross between a translocated strain and a normal strain to be dead (white). How will this manifest itself in the asci? Will you see 50% of your asci containing all dead spores, or all of your asci containing one half-dead and one half-live spores?

Solution

We expect 1/2 asci to segregate N1 + N2/T1 + T2 and from these all spores will be black. We also expect 1/2 segregations to be N1 + T2/ N2 + T1 which will give all aborted (white) ascospores.

6.   Given a cross heterozygous for an inversion in *Neurospora*, you notice in pooled ascospores that 10% are dead.

a.   What patterns of dead and living ascospores do you expect to see if you look at 100 individual asci?

b.   What is the length (in map units) of the inversion?

Solution

a.   20 asci containing 1/2 dead 1/2 live spores, 80 asci with all black spores.

b.   10 m.u.

7. In the fungus *Neurospora*, a standard lab strain carrying the auxotrophic mutation *m* was crossed to a prototrophic strain isolated from a burned sugar cane field. There were three types of asci as shown below, where the symbol '*ab*' means an aborted ascospore whose *m* locus constitution could not be tested.

Type1	Type2	Type3
m	*ab*	*m*
m	*ab*	*m*
m	*ab*	*ab*
m	*ab*	*ab*
+	*ab*	*ab*
+	*ab*	*ab*
+	*ab*	+
+	*ab*	+

Explain the origin of each ascus type.

Solution

This type of pattern suggests heterozygosity for an inversion. First, it is not the all dead/all live pattern expected for a translocation. Second, it seems that all MII patterns have been converted into 50% lethality within those asci suggesting that crossing over is the cause of death. So type 1 is from a meiosis with no crossover, 2 is from a four strand double crossover and 3 is from a single crossover.

8. The ascomycete fungus *Neurospora* has octads. A cross was made between a standard lab strain of genotype *nic-2 leu-3* and a wild-type strain isolated from dung. Previous genetic studies on lab strains showed that *nic-2* is on chromosome 1 and *leu-3* is on chromosome 4. Therefore it was surprising when the nonlinear octad analysis showed only two ascus types as shown below, in equal frequency.

nic	*l eu*	abort
nic	*leu*	abort
nic	*leu*	abort
nic	*leu*	abort
+	+	abort
+	+	abort
+	+	abort
+	+	abort

(Remember that any genome with a deletion or a large inbalance will abort.)

a. Provide a genetic explanation for these results showing clearly the genotypes of the parents under your model.

b. Explain why there are only two ascus types in equal proportions and what they signify.

Solution

   a.   Because of semisterility and because of the all live/all dead pattern suggesting alternate and adjacent segregations, it is likely the cross is heterozygous for a translocation.

   b.   The cross must be *T1 T2* (+ +) x *N1 N2* (*nic leu*) so only *T1 T2/N1 N2* segregations survive. The loci must be both very close to the breakpoint because no recombinant genotypes are observed.

9.   A *Neurospora* inversion spans the region shown in brackets on the following diagram.

$$\longleftarrow 16\ \text{m. u.} \longrightarrow$$

```
_____a_____b_____
          [   inversion   ]
```

A haploid strain with the inversion and of genotype a^+b^+ is crossed to a chromosomally normal strain of genotype *a b*.

   a.   Draw the pairing configuration at meiosis.

   b.   What will be the frequency of a^+b and $a\ b^+$ ascospores?

   c.   What will be the frequency of white (aborted) ascospores?

Solution

   a. There will be a loop   *a*[> loop >]*b*
                              +[< loop <]+

   b.   Since the genes are very close to the breakpoints there will be no viable crossover products.

   c.   Approximately 16% (all normal recombinants converted into lethal genotypes).

10.   In *Drosophila*, *h* determines hairy body (+ = smooth) and *se* determines sepia eyes (+ = red). The loci are approximately 40 m.u. apart on an autosome. A cross was made *se se* + + × + + *hh*, and the F_1 were all wild-type. F_1 females were testcrossed to *se se h h* males and the offspring were

   39   sepia eyes
   42   hairy body

   a.   What progeny and in what proportions were expected?

   b.   Provide an explanation of the observed results if they differ from expectations.

Solution

   a.   Expected: 30% sepia, 30% hairy, 20% wild, 20% sepia hairy.

   b.   The recombinant classes are missing. The most likely explanation is that one of the parental lines was homozygous for an inversion spanning most of the *se–h* region.

11.   In corn, the $+/p$ locus (+ = purple, p = green) is on chromosome 2, and the $+/w$ locus (+ = starchy, w = waxy) is on chromosome 9. A homozygous wild-type was giving unusual results in routine crosses involving these markers. In one analysis it was crossed to green waxy and the F_1 looked wild-type but about half its eggs and pollen aborted. The F_1 was testcrossed and in 50 progeny there were 28 green waxy showing normal gametes, and 22 wild-type showing about 50% abortion of gametes.

    a.   Propose an explanation for these results using diagrams of parents, F_1 and progeny.

    b.   If the F_1 was selfed, what phenotypes would result and in what proportions?

Solution

    a.   The wild-type was homozygous for a translocation involving chromosomes 2 and 9. This explains the semisterility and the presence of only two genoypes in the progeny where four are expected.

        T2 T2 T9 T9 (++ ++) $\times$ N2 N2 N9 N9 (*gg ww*)

        F_1 T2 T9 N2 N9 (+*g* +*w*)

        viable gametes

        T2 T9 (+ +)

        N2 N9 (*g w*)

    b.   1/4 T2 T2 T9 T9 (++ ++)

        1/4 N2 N2 N9 N9 (*gg ww*)

        1/2 N2 T2 N9 T9 (+*g* +*w* semisterile)

12.   In *Drosophila* three mutations were obtained at three separate loci; the dominant mutation star (S), the recessive mutation aristaless (a), and the recessive mutation dumpy (d). All mutations were perfectly viable. A homozygous star strain was crossed to a strain homozygous for both the other mutations. The F_1 females were backcrossed to aristaless dumpy males, and the results were as follows

star	956
aristaless dumpy	918
dumpy	5
aristaless star	7
star dumpy	100
aristaless	132

(The cross showed no evidence of any sterility.)

    a.   Calculate recombinant frequencies and draw a map showing the locations of these loci, with map distances.

    b.   Draw the appropriate chromosomes for the parents and F_1.

    c.   State which phenotypes are missing in the table, and suggest why.

Solution

    a.   The genotypes are *S*++, +*ad*, ++*d*, *Sa*+, *S*+*d* and +*a*+. The missing genotypes are +++ and *Sad*, which must be double recombinants. This fixes the order (using the 'flip' rule) as *a-S-d*. The RFs are 12/2118 for *a* to *S* (= 0.6mu) and 232/2118 for *S* to *d* (11 m.u.).

    b.   Parents must be

a+*d*/*a*+*d* × +*S*+/+*S*+ and F_1 was *a*+*d*/+*S*+.

    c.   The double recombinants are not found because they would be too rare (0.006 × 0.11 × 2118 = about 1).

13.   A pure-breeding strain of corn has the following recessive genes linked in the order shown:

    adl   tes   par   centromere

This plant was fertilized with irradiated pollen from a plant homozygous for all the dominant wild-type alleles. 80 progeny plants were obtained and crossed individually back to the recessive parent. 79 plants gave the following RF values

    adl 10 *tes* 32 *par* 5 cent

although the remaining plant gave the values

    adl 10 *tes* 1 *par* 5 cent.

    a.   What is the probable cause of the abnormal behavior of the single strain?
    b.   How would you test your idea?

Solution

    a.   The crosses are all trihybrid testcrosses. Because the RF between linked genes has been drastically reduced it is likely that the irradiation induced a paracentric inversion in one of the pollen cells treated, and hence the trihybrid is an inversion heterozygote.
    b.   A simple way would be to look for inversion loops and dicentric bridges at meiosis.

9

Chromosome Mutation II: Changes in Number

Multiple-Choice Questions

1. In a mammal how many Barr bodies would be present in cells of individuals who were XXX?
    a.   0
    b.   1
    c.   2 •
    d.   3
    e.   4
    f.   5

2. In a mammal how many Barr bodies would be present in cells of individuals who were XXXY?
    a.   0
    b.   1
    c.   2 •
    d.   3
    e.   4
    f.   5

3.   In a mammal how many Barr bodies would be present in cells of individuals who were XXXXY?
    a.   0
    b.   1
    c.   2
    d.   3 •
    e.   4
    f.   5

4.   In an autotetraploid the genes A/a and B/b are closely linked on either side of the centromere. If a plant of the constitution

    A B
    A B
    a b
    a b

is selfed, and chromosome pairing is as bivalents, what proportion of progeny will be *aaaa bbbb*?
    a.   1/4
    b.   1/1296
    c.   1/36 •
    d.   1/16
    e.   1/1225

5.   How many Barr bodies are found in the nuclei of individuals with Kleinfelter syndrome?
    a.   none
    b.   1 •
    c.   2
    d.   3
    e.   4

6.   Individuals with Turner syndrome
    a.   are triploid.
    b.   are trisomics for chromosome 21.
    c.   are genetic mosaics.
    d.   are monosomic for the X chromosome. •
    e.   are gynandromorphs.

7.   In a tetraploid of genotype *AAaa*, the *A* locus is close to the centromere and pairing is by bivalents. What proportion of gametes will be *Aa*?
    a.   1/8
    b.   1/4
    c.   1/3
    d.   1/2
    e.   2/3 •

8. An allotetraploid is backcrossed to one of its progenitor species and a sterile progeny individual is produced. This sterile individual can be best represented by
   a. $n_1 + n_2$.
   b. $2n_1 + 2n_2$.
   c. $2n_1$.
   d. $2n_2$.
   e. $2n_1 + n_2$. •

9. An allotetraploid can be represented by
   a. $2n_1$.
   b. $2n_2$.
   c. $n_1 + n_2$.
   d. $2n_1 + 2n_2$. •
   e. $n_1 + 2n_2$.

10. In a triploid of genotype *Bbb*, what proportion of gametes will be *B*?
    a. 1/4
    b. 1/3
    c. 1/2
    d. 1/6 •
    e. 2/3

11. The progeny of a cross between a tetraploid and a diploid will be
    a. triploid. •
    b. pentaploid.
    c. diploid.
    d. monoploid.
    e. diploid.

12. The progeny of a cross between a hexaploid and a tetraploid will be
    a. $3n$.
    b. $5n$. •
    c. $2n$.
    d. $7n$.
    e. $10n$.

13. In a tetraploid, x = 5. The ploidy level is
    a. 4. •
    b. 5.
    c. 8.
    d. 10.
    e. 20.

14. Which one of the following terms is not relevant to Down syndrome?
    a. polyploidy •
    b. aneuploidy
    c. first meiotic division nondisjunction
    d. second meiotic division nondisjunction
    e. trisomy 21

15. In *Neurospora*, the genes *his2* and *leu2* are closely linked. From a cross *his2 × leu2* a rare octad of the following type arose:

    hist2 / his2 / his2 / his2 / abort / abort / leu2 / leu2

This octad most likely arose by
    a. first division nondisjunction.
    b. second division nondisjunction. •
    c. post-meiotic mitotic nondisjunction.
    d. crossover between *his2* and *leu2*, followed by first division
    nondisjunction.
    e. rare triploidy.

16. In *Neurospora*, the genes *his2* and *leu2* are closely linked. From a cross *his2 × leu2* a rare octad of the following type arose:

    hist2 / his2 / his2 / his2 / abort / abort / leu2 / leu2

The *leu2* ascospores can be best represented as
    a. *n*.
    b. 2*n*.
    c. *n*-1.
    d. *n*+1. •
    e. 2*n*-1.

17. A human mosaic was discovered of type XXXY/XY. This probably arose by
    a. meiotic nondisjunction in the male parent.
    b. mitotic nondisjunction in a male zygote.
    c. partial polyploidy in a male zygote.
    d. mitotic nondisjunction in a Klinefelter zygote. •
    e. double fertilization.

Open-Ended Questions

1. In the plant genus *Triticum* there are many different polyploid species and also, of course, diploid species. Crosses were made between some different species, and hybrids were obtained. The meiotic pairing was observed in each hybrid, and this is recorded in the following table. (A bivalent is two homologous chromosomes paired at meiosis, and a univalent is an unpaired chromosome at meiosis.)

Species crossed to make hybrid	Pairing in hybrid
1 *T. turgidum* × *T. monococcum*	7 bivalents + 7 univalents
2 *T. aestivum* × *T. monococcum*	7 bivalents + 14 univalents
3 *T. aestivum* × *T. turgidum*	14 bivalents + 7 univalents

Explain these results and in doing so
    a.   deduce the somatic chromosome number of each species used.
    b.   state which species are polyploid, and whether they are auto or allopolyploids.
    c.   account for the chromosome pairing pattern in the three hybrids.

Solution
    a.   1 must be $2n \times 4n$ with one set of 7 in common
           2 might be $2n \times 6n$ with 1 set of 7 in common
           3 must be $4n \times 6n$ with 2 common sets of 7
Putting these together

mono = *AA* ($2x = 14$), *turg* = *AA BB* ($4x = 28$), *aest* = *AA BB CC* ($6x = 42$)

    b.   *turg* and *aest* are allopoyploids
    c.   For example, when *monococcum* is crossed with *turgidum* an *A* gamete combines with an *AB* gamete so that in the hybrid the two *A* sets pair and *B* has no pairing partner.

2.   The $n+1$ female gametophytes (embryo sacs) produced by trisomic plants are usually more viable than the $n+1$ male gametophytes (pollen grains). If 50% of the functional embryo sacs of a selfed trisomic plant are $n+1$ but only 10% of the functional pollen grains are $n+1$, what percentage of the offspring will be
    a.   tetrasomic?
    b.   trisomic?
    c.   diploid?
    d.   nullisomic?

Solution

Male->	$0.9 \ (n)$	$0.1 \ (n + 1)$
Female		
$0.5 \ (n)$	$0.45(2n)$	$0.05 \ (2n + 1)$
$0.5 \ (n + 1)$	$0.45 \ (2n + 1)$	$0.05 \ (2n + 2)$

    a.   0.05
    b.   0.45 + 0.05 = 0.50
    c.   0.45
    d.   0

3.   In a tetraploid plant the *A* and *B* loci are centromere-linked and on seperate chromosomes. A cross is made

$$AAAA\ bbbb \times aaaa\ BBBB$$

and an F_1 is obtained which is selfed to give an F_2. If only one dominant allele is needed to give the dominant phenotypes at each locus, what phenotypes are expected in the F_2 and in what proportions?

Solution

The F_1 is *AAaa BBbb*. The ratios from selfing such a tetraploid are 35:1, therefore

A— B—	$= 35/36 \times 35/36 = 1225/1296$
A— bbbb	$= 35/36 \times 1/36 = 35/1296$
aaaa B—	$= 1/36 \times 35/36 = 35/1296$
aaaa bbbb	$= 1/36 \times 1/36 = 1/1296$

4.   How might a somatic mosaic that had the three genetically different tissues XY, XXY and XXXY?

Solution

The zygote was probably XXY (Klinefelter) and mitotic nondisjunction at some division after the first cell division took two X chromatids to one pole

    <- XX ->
    <-X<-X
    <- YY ->

5.   Assume that you are a genetic counselor. Four phenotypically normal couples who are all expecting their second babies consult you; they want to know about the recurrence risks of various genetic disorders. How do you advise them in each case?

Couple 1.
First child had Down syndrome (DS). No history of any hereditary disorders in their familes. Both aged 25.

Couple 2.
First child had Down syndrome. Several cases of Down syndrome recently in the wife's family.

Couple 3.
First child had Tay-Sachs disease (autosomal recessive).

Couple 4.
First child had achondroplasia (autosomal dominant).

Solution

Couple 1. Most likely DS arose by chance nondisjunction. Risk of a second DS child low.

Couple 2. Most likely wife is heterozygous for a translocation that generates DS. Chance of a second DS child high.

Couple 3. Both must be carriers; chance of second child with Tay-Sachs is 25%.

Couple 4. Because achondroplasia is fully penetrant the child almost certainly arose as a germinal mutation. Recurrence risk depends on the size of the mutant germinal clone; recurrence risk much higher than in population at large.

6. In *Zinnia* plants ($n = 16$), the chromosomes of autotetraploids pair in twos. The locus for flower color (R = red, r = white) is close to the centromere on chromosome 5, and the locus for plant height (D = tall, d = dwarf) is also closely linked to the centromere on chromosome 11. Both dominant alleles show full dominance over any number of recessive alleles.

    a. If you are observing meiosis in tetraploid *Zinnias*, what is the total number of bivalents (chromosome pairs) that will be visible during prophase I? Explain.

    b. In a pollen cell from a tetraploid plant, how many chromosomes will be present?

    c. If a tetraploid plant of genotype *RRrr DDDD* is selfed, what proportion of progeny will be white flowered and tall?

    d. If a tetraploid plant of genotype *Rrrr DDDD* is selfed, what proportion of progeny will be white flowered and tall?

    e. If a tetraploid plant of genotype *RRrr DDdd* is selfed, what proportion of progeny will be white flowered and dwarf? Red flowered and tall?

Solution

    a. Total chromosomes = 64 paired as 32 bivalents.

    b. 32

    c. *rrrr DDDD* = $1/36 \times 1 = 1/36$

    d. *rr* gametes = 1/2, so *rrrr* = 1/4 and *rrrr DDDD* also = 1/4

    e. *rrrr dddd* = $1/36 \times 1/36 = 1/1296$

       $R\text{—} D\text{—} = 35/36 \times 35/36 = 1225/1296 = 0.95$

7. The allotetraploid *Raphanobrassica* is self-fertile but crosses to both parents (cabbage and radish) and produces progeny that are sterile. With the aid of chromosome drawings show why the progeny of these backcrosses are sterile.

Solution

Hybrid is $2n_1 + 2n_2$, which when crossed to $2n_1$ will give $2n_1 + n_2$ hybrid progeny which will have the n_2 set unpaired at meiosis.

8. In tomatoes, the chromosomes of autotetraploids pair in twos at meiosis. Consider a cross *AAaa* × *Aaaa*, in which the A/a gene is close to its centromere. Assuming that only one A allele is needed to produce the A phenotype, what proportion of progeny will be A phenotype? a phenotype?

Solution

Left parent gives 1/6 *aa* gametes. Right parent gives 1/2 *aa* gametes because when A pairs with a remaining pair it must be *aa*. So *aaaa* progeny will be 1/12.

9.   Red-green colorblindness is inherited as an X-linked recessive. A couple with normal vision had a first child who was a normal son and a second child who had Klinefelter syndrome and was also colorblind.
      a.   Draw the relevant chromosomal constitution of these parents and their possible progeny, and explain all the genotypes and phenotypes.
      b.   The couple are planning to have another child: how would you advise them about the possibility of having (i) another Klinefelter child? (ii) another Klinefelter child with colorblindness? (iii) another colorblind child?

Solution
      a.   X-linked locus; woman must have been *Rr*, man *RY*. Daughters will be 1/2 *RR* and 1/2 *Rr*, sons will be 1/2 *RY* and 1/2 *rY*. The first son was obviously *RY*, but the second was *XXY* and must have been *rrY*. Therefore to produce a female gamete that was *rr* nondisjunction must have taken place at the second meiotic division.
      b.   The first Klinefelter child was almost certainly from nondisjunction and the chances of another instance of this (colorblind or not) are negligible. However, there is a 1/2 chance that any sons they might have will be colorblind.

10.   A pair of twins was born, one of which was XYY, and the other was XO. Propose an explanation for this pair of coincident chromosomal anomalies in one pair of twins.

Solution
The zygote was XY. At the first mitotic division of the zygote a mitotic nondisjunction took both Y chromosomes to one pole. The two resulting cells which were XYY and XO became the progenitor cells of a pair of embryos by a process generally responsible for monozygotic twins.

11.   To two parents with no phenotypically visible genetic disorders, a girl is born who has Turner syndrome (designated XO) also hemophilia (an X-linked recessive blood clotting disorder). Propose a full genetic explanation for this coincidence, describing the nature of the event(s) concerned and in which individual and tissue they occurred.

Solution
The mother must have been heterozygous *Hh* and must have contributed an *h* ovum. The man must not have contributed any sex chromosome so there must have been nondisjunction at either the first or second division of meiosis.

12.   Two related species of *Brassica, B. blanca* (2*n* = 12) and *B. japonica* (2*n* = 8) have separate ranges along the coast of Africa. At one place the ranges overlap, and in this location another species *B. neoformans* is found, which 2*n* = 20. *B. neoformans* will not produce fertile offspring in crosses with either *B. blanca* or *B. japonica*. Propose a mechanism for the evolutionary origin of *B. neoformans*, and explain its inability to cross with the other two species.

Solution
A *blanca* × *japonica* cross would give a sterile hybrid of 4 + 6 = 10 chromosomes. If the chromosome set spontaneously doubled, a fertile allotetraploid with 20 chromosomes would arise. When crossed to (say) *blanca*, the progeny would have 4 + 6 + 4 chromosomes, of which six would be unpaired resulting in sterility. Therefore, *neoformans* can cross with the progenitors but no genes can be exchanged making it a new species.

13. In an autotetraploid of genotype *Aaaa*, the locus is close to the centromere.
    a.   If pairing is random by bivalents at meiosis, what proportion of the progeny of a self will be of recessive phenotype?
    b.   How do you think the genotype *Aaaa* was obtained?

Solution
    a.   The *A* chromosome must always pair with a so gametes will be 1/2 *Aa* and 1/2 *aa*. Hence 1/4 progeny will be *aaaa*.
    b.   Make *AAaa* by crossing *AAAA* × *aaaa*. Self this and select *A*— progeny. Self these individually and choose one that gave a progeny ratio of 3/4 *A*—– to 1/4 *aaaa*.

14. In the diploid alpine lily (which has large easily studied chromosomes) the allele *G* causes green leaves and *g* yellow. Two DNA markers M1 and M2 can be detected by simple molecular tests, and act like alleles. The M1/M2 locus is closely linked to *G/g*. A plant of constitution

centromere	G	M1
centromere	g	M2

is grown up in a pot. Although the great majority of the plant tissue is green in appearance (as expected), there was one conspicuous yellow spot on one leaf. Molecular tests on cells from the yellow spot showed that both the M1 and M2 markers could be detected.
    i.   Which of the following genetic hypotheses for the origin of the spot are feasible?
       a.   mitotic crossover
       b.   mutation G->g
       c.   small deletion
       d.   chromosome loss
       e.   mitotic nondisjunction
    ii   How would you eliminate alternative feasible hypotheses?

Solution
i.
    a.   unlikely because this should have rendered the cell homozygous for the M2 allele.
    b.   Possible
    c.   Possible
    d.   Unlikely because M1 and M2 both present.

e.   Unlikely because a first division nondisjunction would give *G* and *g* together and second division nondisjunction would be homozygous for M2. One remote possibility is a crossover between *G/g* and the molecular locus could give a pair of chromatids that is *g* M1/ *g* M2 that would go to the same pole and nondisjoin at second division.

ii.   To distinguish between *b* and *c* take tissue from yellow spot, use to tissue culture to regenerate a whole plant, treat anthers with wide spectrum mutagen, and look for revertants in a cross to *gg* tester. Deletion should show no reversion. It is possible the deletion might produce visible shortening.

15.   In the following pedigree the shaded symbols represent a rare X-linked disease of the blood. The propositus has 45 chromosomes and is sterile. Propose a single mechanism to account for both the sterility and the blood disease in this female.

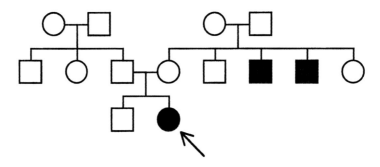

Solution
The woman is probably a case of Turner syndrome (XO). Her mother was a carrier for the recessive disease allele, and in the father's first or second meiotic divisions there was a nondisjunction event giving rise to a sperm lacking an X.

16.   In *Drosophila*, a cross is made involving two closely-linked genes on the very small chromosome 4 as follows, where *a* and *b* are their recessive alleles.

$$\text{Female } \frac{a\ +}{+\ b} \qquad \times \qquad \text{Male } \frac{a\ b}{a\ b}$$

In the progeny, the expected common types were found, but there was also one rare wild-type female.
a.   Explain what the common progeny phenotypes are expected to be, and give their proportions.
b.   Could the rare wild-type female have arisen by crossing over? Explain. By nondisjunction? Explain
c.   The rare wild-type female was test-crossed to a male tester of genotype *aa bb*, and the progeny of the testcross were
1/6 wild-type
1/6 expressed both recessives *a* and *b*
1/3 expressed *b* alone, wild for *a*
1/3 expressed *a* alone, wild for *b*
Which of the explanations in (ii) are compatible with this result? Explain the genotypes and proportions.

Solution

a.   We expect $1/2 \; a +/a \; b$ (*a* phenotype) and $1/2 \; + b/a \; b$ (*b* phenotype).

b.   Yes to both. Rare crossover would give $a + +$ gamete; nondisjunction would give an $a +/+ b \; n + 1$ gamete.

c.   If the wild-type arose by crossover the testcross should have given $1/2 \; + +/a \; b$ and $1/2 \; a \; b/a \; b$ progeny. This is not observed so nondisjunction is likely. In fact the cross progeny ratio fits that expected of a trisomic that is $a +/+ b/a \; b$. This would produce gametes of the following genotypes in equal frequency

$a +/+ b$	wild
$a b$	$a \; b$
$a +/a \; b$	a
$+ b$	b
$+ b/a \; b$	b
$a +$	a

CHAPTER

10

Recombination in Bacteria and Their Viruses

Multiple-Choice Questions

1. Which of the following is not a mode of gene transfer in bacteria?
   a. transformation
   b. specialized transduction
   c. sexduction
   d. lysogeny •
   e. conjugation

2. Bactera belong to the general group of organisms called
   a. auxotrophs.
   b. eukaryotes.
   c. prokaryotes. •
   d. merodiploids.
   e. euploids.

3. A gal mutant
   a. cannot grow without galactose.
   b. is resistant to galactose.
   c. can utilize galactose as a carbon source.
   d. cannot utilize galactose as a carbon source. •
   e. can make its own galactose.

4.   A *strR* mutant
    a.   requires streptomycin.
    b.   can grow in the presence of streptomycin. •
    c.   cannot grow in the presence of streptomycin.
    d.   makes its own streptomycin.
    e.   cannot make its own streptomycin.

5.   The fertility factor is a
    a.   plasmid. •
    b.   virus.
    c.   gene.
    d.   a bacterial mating type.
    e.   bacterial sex hormone.

6.   If a *met thr* F⁺ strain is mated with an F⁻ of genotype *leu thi*, prototrophic recombinants can be detected by plating the mixture on
    a.   leucine and methionine.
    b.   threonine and thiamine.
    c.   leucine and thiamine.
    d.   methionine, threonine, leucine and thiamine.
    e.   minimal medium. •

7.   From an F+ strain with markers in the *c d e b*, which of the following derived Hfrs is impossible
    a.   *c* enters first and *e* last •
    b.   *c* enters first and *d* last
    c.   *d* enters first and *c* last
    d.   *d* enters first and *e* last
    e.   *e* enters first and *b* last

8.   From the following plating data on progeny of the cross Hfr + + + + × F⁻ *his leu met pro* determine the Hfr marker entry sequence (C means colonies detected)

Time(min)	leu met pro	his met pro	his leu pro	his leu met
2	—	—	—	C
4	—	—	C	C
6	C	—	C	C
8	C	C	C	C

In order of entry:
    a.   *leu his pro met*
    b.   *pro met his leu* •
    c.   *his pro met leu*
    d.   *leu pro met his*
    e.   *met leu pro his*

9.   A certain strain behaved as F⁺ *lac⁺* but transferred *lac* ⁺ at high frequency and also threw off F⁻ *lac⁻* cells. Its genotype can best be represented
    a.   F⁺ *lac⁺*
    b.   F⁺ *lac⁻*
    c.   F⁺/F⁻
    d.   F⁺*lac⁺*/F⁺*lac⁻*
    e.   F⁺ *lac⁺*/*lac⁻* •

10.   To demonstrate transformation of bacteria one could
    a.   extract DNA from an auxotroph and add it to prototrophic cells.
    b.   extract DNA from *arg⁻* cells and add it to *arg⁺* cells.
    c.   extract DNA from *str^S* cells and add it to *str^R* cells.
    d.   extract DNA from *arg⁺* cells and add it to *arg* ⁻ cells. •
    e.   mix the DNA from *arg⁺* and *arg⁻* cells to allow recombination.

11.   To demonstrate linkage of two markers A and B by transformation, one needs to demonstrate
    a.   transformation by A = transformation by B.
    b.   transformation by A and B is less than the product of their individual transformation frequencies.
    c.   transformation by A and B is greater than the product of their individual transformation frequencies. •
    d.   transformation by A and B is less than the sum of their individual transformation frequencies.
    e.   transformation by A and B is greater than the sum of their individual transformation frequencies.

12.   The lytic phage gene *h* allows it to infect strains 1 and 2 whereas *H* can infect only strain 1. At another locus *R* causes slow lysis and *r* rapid lysis. A plaque of phenotype cloudy and small will signify infection by
    a.   *H R*. •
    b.   *H r*.
    c.   *h R*.
    d.   *h r*.
    e.   *Hh Rr*.

13.   A virulent phage is always
    a.   lytic. •
    b.   temperate.
    c.   capable of lysogeny.
    d.   capable of producing a prophage.
    e.   capable of zygotic induction.

14.   A bacterial chromosome has four markers A, B, C, and D, evenly spaced throughout the circle. A generalized transducing phage will
    a.   pick up all the markers in one particle.
    b.   only ever transduce one specific marker such as A.
    c.   transduce any of the markers by different transduction events. •
    d.   never transduce more than one marker.
    e.   always transduce at least two of the four markers.

15.   In a generalized transduction experiment using $a + + + +$ donor and an *arg leu pro thr* recipient, pro^+ recipients are selected initially and these are tested further; it is found that 45% have neither arg^+, leu^+ nor thr^+, 48% are leu^+, 6% are $leu^+ thr^+$ and 1% are $leu^+ thr^+ arg^+$. This result shows the gene order is
    a.   *thr - pro - leu - arg.*
    b.   *pro - arg - thr - leu.*
    c.   *pro - leu - thr - arg.* •
    d.   *thr - pro - leu - arg.*
    e.   *pro - arg - thr - leu.*

16.   From one F+ strain the following three Hfr strains were derived, each shown with the first three markers transferred during in an Hfr × F⁻ cross.
    Hfr1 ...DAF->
    Hfr2 ...FBE->
    Hfr3 ...ECD->

The order of genes on the bacterial chromosomal circle must be (A is shown at both ends to represent circularity)
    a.   ADCEBFA. •
    b.   ABCDFEA.
    c.   ACDFEBA.
    d.   AEFBCDA.
    e.   AFBDECA.

Open-Ended Questions

1.   In a cross Hfr $+ + \times$ F⁻ *pho lac*, the *pho* locus enters after *lac*. A sample of pho^+ exconjugants was selected and tested further. It was found that 83% were lac^+ and the remainder were *lac*.
    a.   Explain the origin of the $pho^+ lac^+$, and the $pho^+ lac$ genotypes.
    b.   What do the relative proportions indicate?

Solution

a.

$$\underline{\qquad\overset{+}{\qquad}\qquad\overset{+}{\qquad}\qquad}$$

1	2	3
	pho	*lac*

The + + exconjugants rose from crossovers in regions 1 and 3, and *pho*+ *lac* from crossovers in 1 and 2.

b.   Since there were 17% of the latter type, there must be 17 m.u. between *pho* and *lac*.

2.   A bacterial geneticist has a mixture of cells of various genotypes. He spreads 10,000 cells on a plate containing streptomycin, arginine, and leucine and finds 100 colonies. He uses this as a master plate in a replica plating experiment onto four different media and the results were as follows:

Plate	Additions	Number of colonies
A (control)	*str, leu, arg*	100
B	*str*	25
C	*str, arg*	60
D	*str, leu*	55

What was the number of colonies of genotype

a. str^R leu^+ arg^+
b. str^R leu^+ arg
c. str^R leu arg^+
d. str^R leu arg

Solution

These are all the possible genotypes and all four genotypes can grow on A. However on B only str^R leu^+ arg^+ can grow so there are 25 of these. On C only str^R leu^+ arg^+ and str^R leu^+ arg can grow so there must be 60 - 25 = 35 of the latter. On D only str^R leu^+ arg^+ and str^R leu arg^+ can grow so there must be 55 - 25 = 30 of the latter. The str^R leu arg type can be calculated as 100 - 25 - 35 - 30 = 10.

3 from the cross   Hfr lac^+ pro^+ arg^+ leu^- $str^S \times$
                      F$^-$ lac^- pro^- arg^+ leu^+ str^R

a geneticist wants to obtain progeny of genotype

$$F^- \; lac^+ \; pro^+ \; arg^- \; leu^- \; str^R.$$

How should he go about doing this? (Note: lac^+ = ability to utilize lactose as an energy source, and lac^- = inability to do so.)

Solution

Make the cross, plate exconjugants on streptomycin and lactose but no proline: this selects F$^-$ lac^+ pro^+ str^R. Test a sample of these colonies on a plate lacking arginine and on a plate lacking leucine. Colonies that do not grow on either must be $arg^- leu^-$.

4.   A conjugation experiment is performed in *E. coli* using parents Hfr *arg+ lys+ met+ his+ strS* and F- *arg- lys- met- his_ strR*. Interrupted mating tests were performed at several times and plated on several different media with the following results (the table shows numbers of colonies).

Time (min.)	Medium Streptomycin plus			
	Histidine Lysine Methionine	Arginine Histidine Methionine	Arginine Histidine Lysine	Arginine Lysine Methionine
0	0	0	0	0
10	0	0	0	50
20	20	15	0	400
30	170	200	60	1000
40	600	550	350	1000

a.   What is the order of the four auxotrophic genes with respect to the starting point of transfer (0)?

b.   Conjugation is allowed to proceed for 50 minutes and *met+ strR* cells are selected. A total of 400 *met+* cells are tested further and the following genotypes are obtained:

arg+ lys+   180
arg+ lys-   10
arg- lys+   90
arg- lys-   120

Does this result resolve any uncertainties in your answer to part a? Explain.

Solution

a.   The four media select for *arg+*, *lys+*, *met+* and *his+* respectively. Therefore the order of entry is 0 - *his* - *arg/lys* (can't tell order of these with certainty) - *met*.

b.   Because *met* enters after *arg* and *lys*, this is a standard recombinant frequency analysis. The RF *met-arg* = (90 + 120)/400 = 52.5%. The RF *met-lys* = (10 + 120)/400 = 32.5. Therefore *lys* is closest to *met*.

Note also that the rarest genotype shows that *lys* is in the middle.

5.   Phage P1 was used in a generalized transduction experiment. The genotypes were as follows

donor   *pur1+*   *nadB+*   *pdxJ-*
recipient *pur1*   *nadB_*   *pdxJ+*

Initially transductants of genotype *pur1+* were selected and these were then tested for the other markers. The results were as follows:

nadB+	*pdxJ+*	3 colonies
nadB+	*pdxJ-*	10 colonies
nadB-	*pdxJ+*	24 colonies
nadB-	*pdxJ-*	13 colonies

    a.   What were the cotransduction frequencies of *pur1+* and *nadB+*?
    b.   What were the cotransduction frequencies of *pur1+* and *pdxJ-*?
    c.   What is the order of the three genes?

Solution
    a.   13/50 = 26%
    b.   23/50 = 46%
    c.   Obviously *pdx* is closer to *pur* than is *nad*, but are they both on the same side of *pur* or on opposite sides? Note that of the 13 *nad+* colonies, most (10) have the *pdxJ-* allele, suggesting that they are transduced together and that the order is *pur-pdx-nad*. (By comparison, note that of the 23 *pdxJ-* colonies less than half have *nadB+*.) This order is confirmed by the fact that this is the order which produces one rare genotype resulting from two double crossovers (the *pur1+ nad+ pdxJ+* genotype).

6.   An *E. coli* strain of genotype *pur+ pdxJ+ glyA-* was infected with the generalized transducing phage P1 and the progeny phages were used to transduce a strain of *pur1- pdxJ- glyA+*. The selection was made for *pur1+* and then the transductants were tested for the other markers. The results were:

pdxJ+	*glyA-*	5
pdxJ+	*glyA+*	18
pdxJ-	*glyA-*	19
pdxJ-	*glyA+*	8

    a.   What was the cotransduction frequency of *pur1* and *pdxJ*?
    b.   What was the cotransduction frequency of *pur1* and *glyA*?
    c.   What was the order of the three genes? (Explain your logic.)
    d.   The following formula converts cotransduction frequency into map units corresponding to minutes. L is the mean size of the P1 transducing fragment, and equals 2 minutes. d is the map distance in minutes.
    cotransduction frequency = $(1 - d/L)^3$
Calculate the map distances between *pur* and *pdx*, and between *pur* and *gly*.

Solution
    a.   23/50 = 0.46
    b.   24/50 = 0.48

c. *pdx* and *gly* are both the same distance from *pur*, but, they must be on opposite sides. If they were on the same side then one donor allele should nearly always be accompanied by the other donor allele and this is not shown. Also the genotype *pur⁺ pdx]⁺ glyA⁻* should be one of the more common genotypes.

d. *pur - pdx* = 0.46 min., *pur - gly* = 0.43 min.

7. DNA from a *Bacillus subtilis* strain of genotype *gua⁺ pac⁻ dna⁺* was used to transform a recipient strain of genotype *gua⁻ pac⁺ dna⁻* (Note: *dna* is a DNA polymerase gene.) Transformants of genotype *gua⁺* were selected and then tested for the other markers. Out of 10 transformants the distribution was as follows

pac⁻	*dna⁺*	6
pac⁻	*dna⁻*	1
pac⁺	*dna⁺*	44
pac⁺	*dna⁻*	56

a. What is the cotransformation frequency of *gua* and *pac*?
b. What is the cotransformation frequency of *gua* and *dna*?
c. What is the most likely gene order? (Draw a rough map giving some idea of relative distances.)

Solution

a. 7/107 = 6.5%
b. 50/107 = 47%
c. Most likely *gua* - (close) - *dna* - (distant) - *pac*, because most *pac⁻* transformants are *dna⁺* (6/7). This order requires that *gua⁺ dna⁻ pac⁻* be produced by 2 double crossovers, explaining the rarity of this type.

8. In a certain Hfr the transfer of genes is in the following order:

ser *met* *leu* *arg*

————————————————————>

An Hfr: *arg⁺ leu⁺ met⁻ ser⁺* and an F⁻: *arg⁻ leu⁻ met⁺ ser⁻* were crossed and plated on media that selected for *ser⁺* exconjugants. When these *ser⁺* colonies were tested for the unselected markers the following was found:

Selected	Unselected
ser⁺	74% *leu⁺* / 26% *leu⁻*
ser⁺	59% *arg⁺* / 41% *arg⁻*
ser⁺	12% *met⁺* / 78% *met⁻*

a. What is the map distance between *ser* and *leu*?
b. What is the map distance between *leu* and *arg*?

    c.   What is the map distance between *met* and *ser*?

    d.   Another gene (*pur*) has been mapped near *leu*. The above Hfr is *pur⁻* and the F⁻ recipient is *pur⁺*. When some *ser⁺* exconjugants were plated on complete media minus *leu* many colonies were obtained. When some *ser⁺* exconjugants were plated on complete media minus *pur* and *leu* many colonies were still obtained. On which side of *leu* is the *pur* locus? Show below both possible locations and why each is correct or incorrect.

Solution

    a.   RF *ser - leu* = 26%.

    b.   RF *ser - arg* = 41%, therefore RF *leu - arg* = 15% by subtraction.

    c.   RF *ser - met* = 12%.

    d.   If *pur⁻* had been to the left of *leu⁺*, then selecting for *leu⁺* would have required two double crossovers, so *pur* is probably to the right. This fits the second selection in which the second crossover must occur between *leu* and *pur* to pick up both + alleles.

9.   A transformation was performed with *arg⁺ gly⁺* donor DNA and *arg⁻ gly⁻* recipient cells. The results are shown below.

Transformed genotype	% cells transformed
arg⁺ gly⁻	28
arg⁻ gly⁺	43
arg⁺ gly⁺	12

Are these two genes closely linked? Explain.

Solution

Double transformants are found at a frequency that is the product of the two single transformation frequencies ($0.28 \times 0.43 = 0.12$) therefore the two genes are very distant from each other, so distant that they can never be both included on a tranforming fragment.

10.   Four strains of *E. coli* were being used for conjugation experiments. Strains A and B were Hfr, and C and D were F_. The results were as follows

    A × C   donor genes transferrred

    B × D   donor genes transferred

    A × D   lysis

    B × C   donor genes transferred

    a.   What might be unique about the third cross?

    b.   Why do the same strains A and D not show lysis in other crosses?

Solution
- a. Zygotic induction seems to be occurring.
- b. A and C both have a lysogenized phage, but B and D do not.

11. In a certain Hfr the transfer of genes is in he following order:

—*ser*——*met*——*leu*——*arg*——>

An cross was made Hfr *arg*+ *leu*+ *ser*+ × F- *leu*- *met*- *ser*- and cells were plated on medium that selected for *met*+ exconjugants. When these were tested for unselected markers the following frequencies were found:

Selected	Unselected	
met+	72% *leu*+	28% *leu*-
met+	56% *arg*+	44% *arg*-
met+	32% *ser*+	68% *ser*-

- a. What is the map distance between *met* and *leu*?
- b. What is the map distance between *leu* and *arg*?
- c. What is the map distance between *met* and *ser*?

Solution
- a. 28 map units
- b. *met - arg* = 44, *leu - arg* = 44 - 28 = 16
- c. Cannot be determined because *ser* enters after *met*.

C H A P T E R

11

The Structure of DNA

Multiple-Choice Questions

1. A sample of normal double-helical DNA was found to have a guanine content of 18%. What is the expected proportion of adenine?
    a. 9%
    b. 32% •
    c. 36%
    d. 68%
    e. 82%

2. A gene is
    a. a transcriptional unit of DNA. •
    b. any segment of DNA.
    c. the segment of DNA between the initiator and stop codons.
    d. the segment of DNA corresponding exactly to the mRNA, tRNA or rRNA.
    e. any region of DNA coding for a polypeptide.

3. Which of the following <u>is</u> a component of DNA
    a. cytosine •
    b. arginine
    c. guanidinium
    d. tyrosine
    e. alanine

4.   In the division of a somatic cell, DNA replication occurs
     a.   at prophase.
     b.   at metaphase.
     c.   at anaphase.
     d.   at telophase.
     e.   before mitosis. •

5.   In DNA, the bond between a deoxyribose sugar and phosphate is a
     a.   polar bond.
     b.   ionic bond.
     c.   hydrogen bond.
     d.   covalent bond. •
     e.   Van der Waal's force.

6.   In a normal human body cell, how many DNA molecules are there in the nucleus?
     a.   many thousands
     b.   1
     c.   23
     d.   46 •
     e.   92

7.   In one strand of DNA the nucleotide sequence is 5′ ATGC 3′. The complementary
sequence in the other strand must be
     a.   5′ CGTA 3′.
     b.   3′ ATGC 5′.
     c.   5′ TACG 3′.
     d.   5′ ATGC 3′.
     e.   3′ TACG 5′. •

8.   How many different DNA molecules eight nucleotide pairs long are thoretically
possible?
     a.   24
     b.   64
     c.   65,536 •
     d.   256
     e.   32

9.   If phage are labelled with radioactive sulfur and allowed to infect bacterial cells,
the phage progeny resulting from lysis are expected
     a.   to be non-radioactive. •
     b.   to have radioactive DNA.
     c.   to have radioactive proteins.
     d.   to have radioactive DNA and proteins.
     e.   to have radioactive carbohydrates.

10. Okazaki fragments form on the
    a.   mRNA.
    b.   the 3′ end of a polymerizing strand of DNA.
    c.   the leading strand.
    d.   the lagging strand. •
    e.   major groove of DNA.

11. If a 1000 kilobase fragment of DNA has 10 evenly spaced replication origins and DNA polymerase moves at 1 kilobase per second, how many seconds will it take to produce two daughter molecules?
    a.   100
    b.   20
    c.   30
    d.   40
    e.   50 •
(Note to instructor: each origin replicates bidirectionally.)

12. If radioactive sulfur is used to label phages before infection of bacteria, after lysis the radioactivity will appear predominantly in the
    a.   bacterial DNA.
    b.   progeny phage DNA.
    c.   bacterial proteins.
    d.   progeny phage proteins.
    e.   phage ghosts. •

13. In double-stranded DNAs generally, which of the following is true?
    a. $C = T$
    b. $A = G$ and $C = T$
    c. $A + G = C + T$ •
    d. $A = C$ and $G = T$
    e. $A = G$

14. The <u>backbone</u> of a DNA strand is composed of
    a.   sugars and phosphates .•
    b.   sugars and bases.
    c.   bases and phosphates.
    d.   bases and nucleotides.
    e.   sugars and nucleotides.

15. Which of the following is not true for double stranded DNAs generally
    a. $A = T$
    b. $G = C$
    c. $C = T$ •
    d. $A + G = T + C$
    e. $A + C = G + T$

16. Which of the following is not found in DNA?
    a.   phosphorus
    b.   nitrogen
    c.   hydrogen
    d.   oxygen
    e.   sulfur •

17. When does DNA replication occur in relation to mitosis?
    a.   prophase
    b.   metaphase
    c.   anaphase
    d.   telophase
    e.   before mitosis •

Open-Ended Questions

1.   It is estimated that a human genome is 3 billion nucleotide pairs, and that it contains 100,000 genes. How many nucleotides of DNA is occupied by an average gene plus intergenic sequence?

Solution
$3 \times 10^9 / 10^5 = 3 \times 10^4 = 30$ kilobase pairs (30 kb).

2.   Assume that a DNA molecule has one replication origin exactly in the middle of its length. Draw the DNA at a time when it is half replicated. Include representations of DNA polymerase, leading and lagging strands, and label all 5′ and 3′ ends.

Solution

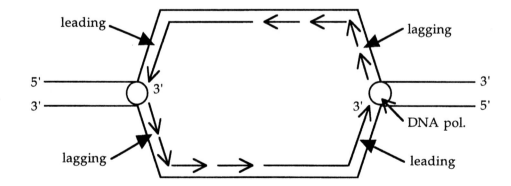

3.   A fragment of double helical DNA is found to have 100,000 nucleotide pairs.
    a.   What is the total number of nitrogen bases in this fragment?
    b.   What is the total number of phosphate atoms in this fragment?

Solution
    a.   200,000
    b.   200,000

4.   In a certain double helical DNA molecule the ratio of A + G:T + C in one strand is 0.4.
    a.   What must be the A + G:T + C ratio on the other strand?
    b.   What must be the ratio of A + G:T + C in the whole double-stranded molecule?

Solution
    a.   Since A pairs with T and G with C, the ratio in the other strand is the inverse of that in the strand given; that is, instead of 0.4:1 it must be 1:0.4 or 2.5:1
    b.   1.0

5.   In a certain double helical DNA molecule the A + T:C + G ratio in one strand is 0.4.
    a.   What must be the A + T:C + G ratio in the other strand?
    b.   What must be the A + T:C + G ratio in the entire molecule?

Solution
    a.   0.4
    b.   0.4

6.   Adenine-requiring auxotrophic bacterial cells are grown in the presence of adenine containing the heavy isotope of nitrogen (^{15}N) for many cell division cycles until all DNA is heavy. The cells are then grown in the normal 'light' isotope (^{14}N) for one and two cell divisions. The DNA is centrifuged in cesium chloride density gradients. How many bands and in what relative positions would be shown by DNA
    a.   from the original cells?
    b.   cells after one division?
    c.   cells after two divisions?

Solution
    a.   One band at low position in gradient
    b.   One band at a higher position
    c.   Two bands, on at position from b, and one even higher.

7.   Cells are grown in light nitrogen (N^{14}) until all the nitrogen in their DNA is light. Then they are grown for two mitotic division cycles in heavy nitrogen (N^{15}). Finally, they are switched back to light nitrogen for one division. Using solid lines for heavy nitrogen, and dotted lines for heavy nitrogen-containing polynucleotide strands, sketch the DNA at these four stages. (If there are mixtures at any stage, be sure to give the proportions of types.)

Solution

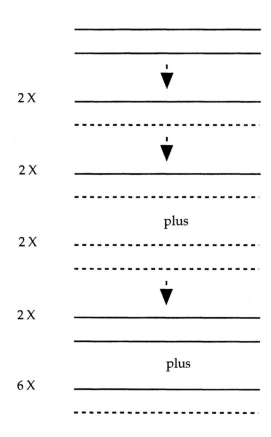

8. Some species in the diploid plant genus *Happlopappus* have extremely small chromosome numbers. Assume that *H. gracilis* has only one pair of homologous chromosomes, in other words $2n = 2$. A specific plant is heterozygous (*Aa*) for a gene very closely linked to the centromere. Draw labelled schematic diagrams that show the chromosomes at the following stages

Mitosis
  a. just before the mitotic S phase
  b. mitotic metaphase
  c. mitotic anaphase

Meiosis
  d. just before premeiotic S phase
  e. metaphase of the first meiotic division
  f. anaphase of the first meiotic division
  g. anaphase of the second meiotic division

Then repeat a through g representing chromosomes and chromatids as DNA. (Throughout this question, when you use a line in your drawings, make sure you define what that line represents (chromosome?, chromatid?, double helix?, polynucleotide strand?, etc.)

Solution

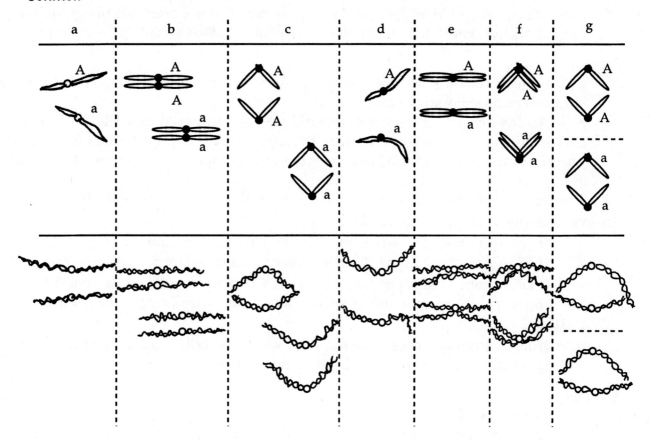

9.   Diploid plant cells dividing in culture medium (nutritive solution) were fed radioactive thymine nucleotides for the duration of the premitotic S phase only, and the radioactivity was incorporated into growing polynucleotide chains. The cells were then returned immediately to nonradioactive medium. In the mitotic prophase following exposure, the chromosomes were tested for radioactivity using a technique called autoradiography. In autoradiography a chromatid that has radioactivity in one or both DNA polynucleotide strands is seen with spots all over it.

   a.   The two following diagrams represent a pair of homologous chromosomes divided into chromatids at mitotic prophase following exposure. Which diagram (A or B) best represents the expected radioactivity patterns? Explain your answer with diagrams of DNA.

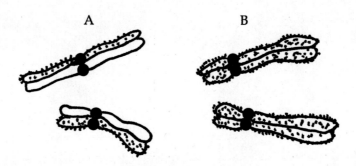

b.   Using the same types of chromatid diagrams, sketch the pattern of radioactivity expected in the prophase of the next mitosis after this one if kept in nonradioactive medium. Explain your answer at the DNA level.

Solution
a.   Both chromatids will be labeled because each consists of a DNA molecule with one radioactive strand and one not.
b.   At this division one sister chromatid will be composed of a DNA molecule that is entirely nonradioactive, and the other sister will be half radioactive. Therefore one chromatid will have no dots and the other will be covered with dots.

10.   Proline is normally synthesized by bacterial cells but proline-requiring mutants are available (*pro*⁻). What is the outcome?
a.   If 10^6 cells from a *pro*⁻ strain are spread on minimal medium?
b.   If DNA from a wild-type strain is spread on minimal medium?
c.   If DNA from a wild-type strain is mixed with 10^6 *pro*⁻ cells whose membranes have been made porous and then the whole mixture is plated on minimal medium?
d.   If DNA from *pro*⁻ cells is mixed with wild-type cells whose membranes are made porous, and the mixture is plated on minimal medium?

Solution
a.   no colonies; perhaps a few revertants.
b.   nothing; DNA can't grow. (However if some wild-type cells are contaminating the DNA solution then some colonies might appear. Check another marker from the recipient strain.)
c.   Lots of colonies from transformation.
d.   The whole plate would be covered with growth; the *pro*⁻ DNA wouldn't have any effect — if transformation to *pro*⁻ occurred these cells would be swamped.

C H A P T E R

12

The Nature of the Gene

Multiple-Choice Questions

1. In *Neurospora*, a linear biochemical pathway synthesizes an amino acid D

    E1 E2 E3
    A -> B -> C -> D

Null mutants for the enzyme E2 gene will grow on minimal medium supplemented
with compounds
    a. A or B.
    b. C or D. •
    c. A or B or C or D.
    d. D only.
    e. A only.

2. A branched biochemical pathway synthesizes two related essential amino acids D
and F

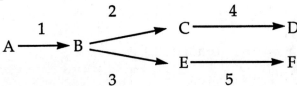

A mutant defective for enzyme 2 will grow on minimal medium supplemented with
    a. A or B.
    b. E or F.
    c. C or F.
    d. D or F.
    e. C or D. •

3.   A branched biochemical pathway in a bacterium synthesizes two related essential amino acids D and F

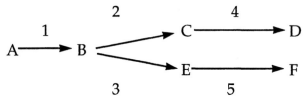

A mutant defective for enzyme 1 will grow on minimal medim plus
   a.   A alone.
   b.   B alone. •
   c.   C alone.
   d.   C and D.
   e.   E and F.

4.   In a fungus, two biochemical pathways converge to make essential substance D:

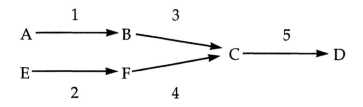

In heterokaryons, complementation will occur between mutants in genes for enzymes
   a.   1 and 2.
   b.   1 and 3.
   c.   2 and 3.
   d.   3 and 4.
   e.   All of the above. •

5.   In a plant, two pathways combine to produce an orange color in the wild-type

```
           Enz 1
white ───────────────▶ yellow  ⎫
                               ⎬ =   orange
white ───────────────▶ red     ⎭
           Enz 2
```

If a mutant homozygous for a null mutation in the gene for enzyme 1 is crossed to a plant homozygous for a null mutation in the gene for enzyme 2, the F_1 will be
   a.   all white.
   b.   all yellow.
   c.   all red.
   d.   all orange. •
   e.   9:3:3:1 for red:orange:yellow:white.

6. In a plant, two pathways combine to produce an orange color in the wild-type

white $\xrightarrow{\text{Enz 1}}$ yellow

white $\xrightarrow[\text{Enz 2}]{}$ red

$\left.\begin{array}{c} \\ \\ \end{array}\right\}$ = orange

If a mutant homozygous for a null mutation in the gene for enzyme 1 is crossed to a plant homozygous for a null mutation in the gene for enzyme 2, the F_2 will be

a.   9 orange:3 yellow:3 red:1 white •
b.   9 orange:7 white
c.   9 orange:4 red:3 yellow
d.   12 orange:3 yellow:1 red
e.   9 orange:4 yellow:3 red

7.   In a bacterium, mutants unable to synthesize an essential compound G were tested with related compounds D, E, and F. The results were as follows where + means growth.

Compounds Mutants	D	E	F	G
1		+		+
2		+	+	+
3				+
4	+	+	+	+

The order of gene action in the biosynthetic pathway for G must be

a.   3 -> 4 -> 1 -> 2
b.   1 -> 4 -> 3 -> 2
c.   1 -> 2 -> 3 -> 4
d.   4 -> 2 -> 1 -> 3 •
e.   4 -> 3 -> 2 ->1

8.   In yeast, four short viable deletions (1–4) encompassing the *ade1* locus were intercrossed and the ascospore plated on minimal medium to determine if there were any wild-type prototrophic recombinants. The results were as follows where + means that wild-type recombinants were produced

	1	2	3	4
1		+		
2				+
3	+			+
4		+	+	

The order of the deletions must be
   a.   3 2 1 4 •
   b.   2 4 1 3
   c.   1 3 4 2
   d.   4 1 3 2
   e.   3 1 2 4

9.   In *Drosophila*, a recessive point mutation *m* is paired with each of four overlapping viable deletions.

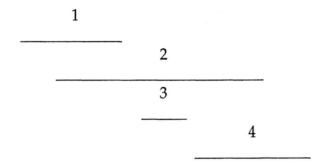

The *m* phenotype was not expressed in combination with deletions 1 or 3 but it was in combination with 2 or 4 . Therefore, *m* must be located in the
   a.   left portion of 1.
   b.   1/2 overlap.
   c.   2/3 overlap.
   d.   2/4 overlap. •
   e.   right hand portion of 4.

10.   In yeast, an arginine gene *argR* is very closely linked to genes *M* and *N*. A cross was made *M N argR1* × *m n argR2*, where *argR1* and *argR2* are separately obtained mutant alleles of the *argR* gene. Ascospores were plated on medium containing no arginine but supplemented for *M* and *N*. The *argR*+ prototrophs that grew were tested further and were found to be all of genotype *m N*. The gene and mutant site order must be
   a.   *M-N-1-2.*
   b.   *1-2-M-N.*
   c.   *M-2-1-N.*
   d.   *M-1-2-N.* •
   e.   *M-N-2-1.*

11.   Mutations 1,2, and 3 of the *Neurospora* gene *cys-1* are intercrossed and the frequency of prototrophic recombinants was

$1 \times 2 : 0.6$ per 10^4

$1 \times 3 : 0.7$ per 10^4

$2 \times 3 : 1.0$ per 10^5

The mutant site in the middle position must be
   a.   1.
   b.   2. •
   c.   3.
   d.   1 or 3 (can't tell).
   e.   2 or 3 (can't tell).

12.   In *Neurospora*, a complementation test (using heterokaryons) was performed on five haploid mutant strains with abnormal hyphal branching. The results were as follows where + means that the heterokaryon was wild-type in appearance, and - means abnormal branching phenotype.

	1	2	3	4	5
1	−	+	+	−	−
2	+	−	+	+	+
3	+	+	−	+	+
4	−	+	+	−	−
5	−	+	+	−	−

These results indicate that the mutations were in
   a.   1 gene.
   b.   2 genes.
   c.   3 genes. •
   d.   4 genes.
   e.   5 genes.

13.   In *Neurospora*, a complementation test (using heterokaryons) was performed on five haploid mutant strains with abnormal hyphal branching. The results were as follows where + means that the heterokaryon was wild-type in appearance, and - means abnormal branching phenotype.

	1	2	3	4	5
1	−	+	+	−	−
2	+	−	+	+	+
3	+	+	−	+	+
4	−	+	+	−	−
5	−	+	+	−	−

The following mutations are all in one gene.
   a.   1, 3, 5
   b.   2, 3, 5
   c.   3, 4, 5
   d.   2, 3, 4
   e.   1, 4, 5 •

14. Which of the following is not a genetic disease?
     a.   phenylketonuria
     b.   cystic fibrosis
     c.   Tay-Sachs disease
     d.   tuberculosis •
     e.   Huntington's disease

15. If a heterokaryon is made between mutant strains caused by mutations in different genes in adenine biosynthesis, the phenotype of the heterokaryon will be
     a.   lethal.
     b.   wild-type. •
     c.   the same as the two original strains.
     d.   adenine-requiring.
     e.   more severely adenine requiring.

16. To select for heterokaryons after mixing the following two strains
     ad-1 nic-4 his-2
     leu-3 nic-4 met-2

the medium should be
     a.   minimal medium (mm).
     b.   mm + adenine, nicotinamide, and histidine.
     c.   mm + adenine, histidine, leucine and methionine.
     d.   mm + adenine, nicotinamide, histidine, leucine and methionine.
     e.   mm + nicotinamide. •

17. A biochemical pathway making pigments shows the following sequential color conversions, each catalyzed by seperate enzymes A, B, and C

        colorless->yellow->blue->red
                 A   B   C

A null mutation in the B gene will result in a phenotype that is
     a.   orange.
     b.   purple.
     c.   yellow. •
     d.   green.
     e.   white.

18. A biochemical pathway making pigments shows the following sequential color conversions, each catalyzed by seperate enzymes A, B, and C

        colorless->yellow->blue->red
                 A   B   C

A null mutation in each of the genes A and B will result in a phenotype that is
    a.   orange.
    b.   purple.
    c.   yellow.
    d.   green.
    e.   white. •

19.   A plant biochemical pathway making pigments shows the following sequential color conversions, each catalyzed by seperate enzymes A, B, and C

colorless->yellow->blue->red
         A   B   C

If a plant homozygous for a null mutation in the B gene is crossed to a plant homozygous for a null mutation in the C gene, the F_2 will show the following phenotypic ratio
    a.   9 red:7 blue
    b.   9 red: 3 blue:3 yellow:1 white
    c.   13 red: 4 blue
    d.   9 red: 3 blue: 4 yellow •
    e.   12 red: 3 blue: 1 yellow

Open-Ended Questions

1.   Two different *rII* mutant strains were used to infect *E. coli* B. Bacteria lyse. The resulting phages were used to singly infect *E. coli* K. and plaques were found; are these plaques on *E. coli* K most likely produced by complementation or recombination?

Solution
Recombination. Since only a single phage attacks each K cell, that phage must be wild-type: there is no opportunity for complementation to occur.

2.   In an *E. coli* cross the str^R Hfr strain carried a nonrevertable allele of the *trp1* gene, and the str^S F⁻ carried a different nonrevertable allele of the same *trp1* gene. Exconjugants were plated on medium containing streptomycin but lacking tryptophan; surprisingly a small number of colonies was found. How would you explain such colonies?

Solution
Str^R was obtained from the Hfr. Furthermore ther must have been a double crossover of which one of the crossovers occurred between the mutant sites of the *trp1* gene. Since this region is very small, few such recombinants were observed.

3.   Cells of two strains of *Neurospora* (one was *ad-3 pan-2* and the other was *leu-3 met-6*; all alleles conferring auxotrophic requirements) were mixed, suspended in water, pelleted by centrifugation, and the pellet was placed on minimal medium. A large vigorous colony grew. Control experiments showed that the two original strains did not grow on this medium, Explain how the colony was able to grow vigorously on minimal medium.

Solution
The two cell types fused to form a heterokaryon (both nucler types in a common cytoplasm); complementation occurred because the heterokaryon has the wild-type alleles for all loci.

4.   In the synthesis of the amino acid tryptophan in *E. coli* a number of different mutations were induced by a chemical that causes a mixture of deletions and point mutations. These all mapped to one locus so were presumed to affect one gene. The mutant stocks were mated through transduction, and the recipients were plated to determine if there were any wild-type recombinants. The following grid shows the results: + means recombinats were detected, 0 means that none were

	1	2	3	4
1	0	0	0	+
2	0	0	+	0
3	0	+	0	+
4	+	0	+	0

a.   Considering that all or most might be deletions, draw a deletion map from these data.
b.   A known point mutation for this gene gave wild-type recombinants with the above strains 3 and 4, but not with 1 or 2. At which position on the deletion map is the mutant site of this mutation located?

Solution
    a.

                                        4
                                      ___
                              2
                      _____
              1
        _____
      3
  _____
          ___

b.   Mutant site (asterisked)
      is here   *

5. In a mutant hunt, ten adenine-requiring mutations of yeast were obtained and a complementation test revealed the following results

	1	2	3	4	5	6	7	8	9	10
1	−	−	+	+	+	+	−	+	+	+
2		−	+	+	+	+	−	+	+	+
3			−	+	−	+	+	−	+	+
4				−	+	−	+	+	−	−
5					−	+	+	−	+	+
6						−	+	+	−	−
7							−	+	+	+
8								−	+	+
9									−	−
10										−

a. Interpret these results.

Mutants 1, 3, and 4 were tested for growth on the compounds CAIR, AIR and SAICAR, all related to adenine chemically. The results were

	Adenine	CAIR	AIR	SAICAR
1	+	−	−	+
3	+	−	−	−
4	+	+	−	+

b. Explain these results.

Mutants 1, 3, and 4 were intercrossed and 1000 ascospores were plated from each cross onto minimal medium to select for prototrophs. The results were as follows

	1	3	4
1	0	4	245
3	7	0	260
4	255	252	0

c. Explain these results and integrate them with previous results to provide a summary statement about this genetic system.

Solution

a. Three complementation groups (genes) containing
1, 2, 7/3, 5, 8/4, 6, 9, 10   (can't tell order)

b. AIR——4——>CAIR——1——>SAICAR——3——>ADENINE

c. Genes containing 1 and 3 very closely linked (few ++ recombinants), but gene containing 4 is unlinked (25% ++ = 25 × 2 = 50% RF).

6.   A *Drosophila* mutant search was carried out to find recessive lethal mutations on chromosome 2. Ten mutations of this type were obtained, and from these stocks hybrid zygotes were produced that bore the lethal chromosome from one stock, plus the lethal chromosome from the other. In other words, they could all be represented symbolically as l_1/l_2. All pairwise combinations were tested to find out if the zygotes were viable or not. The results were as follows, where a plus means the zygotes were viable, and a minus means the zygotes were not viable.

	1	2	3	4	5	6	7	8	9	10
1	−	+	−	+	+	+	−	+	+	+
2		−	+	+	+	+	+	+	+	−
3			−	+	+	+	−	+	+	+
4				−	+	−	+	+	+	−
5					−	+	+	+	−	−
6						−	+	+	+	−
7							−	+	+	+
8								−	+	+
9									−	−
10										−

Interpret these results as fully as possible showing
  a.   How a + and a − result can be produced genetically
  b.   How many genes are involved in generating these results
  c.   What can you say about their location?
  d.   The meaning of any anomalous type(s)

Solution
  a.   A − shows the mutations are alleles therefore the genotype is homozygous lethal. A + means nonalleles.
  b.   Five genes $1, 3, 7/2/4, 6/5, 9/8$
  c/d. Some information about order and linkage can be obtained by considering the observation that 10 must be a deletion because it fails to complement the three genes shown above as the 'inner' three. However, order within the area spanned by the deletion cannot be deduced, neither can the order of the two genes outside the deletion.

7.   In *Drosophila*, the lozenge locus is on the X chromosome. Recessive lozenge mutations (*lz*) make the eye oval shaped instead of the usual round shape. Furthermore, it is known that when two mutations occur within the lozenge gene, the phenotype is extreme lozenge (more oval than one lozenge mutation).

The genes *sn* and *v* are closely linked to either side of *lz*. Three single mutations, *lz3*, *lz4* and *lz7*, were obtained and used in the following experiments.

i.   In 12,000 progeny of a female of genotype *sn lz4 v/ + lz7 +*, most males were regular lozenge phenotype, but two males were phenotypically extreme lozenge and of *v* phenotype, and one other male had wild-type eyes and was of *sn* phenotype.

ii.   In 14,000 progeny of a female of genotype *sn lz3 v/ + lz7 +*, most males again were of regular lozenge appearance, but one male had wild-type eyes and was *v* phenotype, and three males had extreme lozenge eyes and were of *sn* phenotype.

a.   Draw gene diagrams that illustrate the difference between wild-type, lozenge, and extreme lozenge.
b.   Illustrate the origin of the rare males with the aid of diagrams.
c.   Explain why the rare males are rare.
d.   What do the results tell us about the relative positions of the mutant sites at the lozenge locus in relation to *sn* and *v*?
e.   In the progeny of a *+ lz3 v / sn lz4 +* female, what should be the *sn* and *v* phenotypes of rare extreme lozenge males and males with wild-type eyes?

Solution

a.   wild   _____
      lozenge   _____*_____
      lozenge   _____*_____
      extreme   _____*_____*_____

b.   Crossovers between the mutant sites.
c.   The mutant sites are close so crossovers there are rare.
d.   *sn*-3-4-7-*v*
e.   The cross is +—3+—*v/s*—+4—+, so wild-types will be *sn v* and extremes will be ++.

8.   In haploid yeast, a large number of tryptophan-requiring mutants was obtained. These were tested to determine if other compounds related to tryptophan would allow growth. The mutants fell into four groups as follows, where + means growth and - means no growth.

Supplement	Mutant group			
	1	2	3	4
dihydroshikimic acid	–	+	–	+
protocatechuic acid	–	–	–	+
shikimic acid	+	+	–	+
trytophan (control)	+	+	+	+

a.   What do these data tell us about the normal action of the wild-type alleles corresponding to each of these mutant groups in the synthesis of tryptophan?

Mutants in each of the groups were intercrossed, and the recombinant frequencies as percentages were found to be as follows

	1	2	3	4
1	0	0.1	50	50
2		0	50	50
3			0	50
4				0

  b. What do these results tell us? (Draw a diagram if appropriate.)

Solution

  a. Biosynthesis is as follows

  —4—> proto —2—> dihyd —1—> shik —3—> tryp

  b. 1 and 2 closely linked, 3 and 4 on different chromosomes

  _____1_2_____   ____3____   ____4____

9. Starting with a haploid wild-type strain of yeast (fully prototrophic), a mutational screen obtained large numbers of mutants that were auxotrophic for the amino acid histidine (that is, required histidine for growth). These mutants were tested for their ability to grow on chemicals structurally related to histidine, and it was found that they could be grouped into four classes 1 – 4. The following diagram illustrates the responses in each class by using one representative strain from each class. (The big circles represent a growing colony of cells originating from small inoculum, and the asterisks represent inocula which failed to grow. The labels indicate what was added to the plates.)

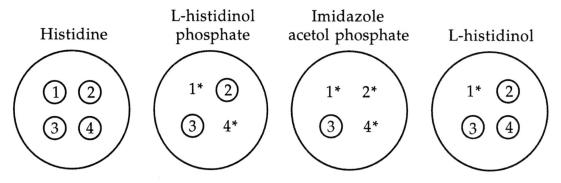

  a. Propose a biochemical pathway for the synthesis of histidine.
  b. Show the relationship between the steps in the pathway and the genes in the four groups of mutants.

Solution

  a and b. The mutations are in four separate genes coding for enzymes working sequentially in a biochemical pathway synthesizing histidine.

precursor-3->imidaz.-2->his. phos.-4->histidinol-1->histidine

10. In humans, PKU (phenylketonuria) is a disease caused by an enzyme deficiency in step A in the following simplified reaction sequence, and AKU (alkaptonuria) is due to an enzme deficiency in step B.

$$\text{phenylalanine} \xrightarrow[A]{} \text{tyrosine} \xrightarrow[B]{} \text{carbon dioxide}$$

A person with PKU marries a person with AKU:
   a.   designate genotypes of the two individuals using symbols of your own choosing (be sure to define them).
   b.   what will be the genotype(s) of their children?
   c.   what will be the phenotype(s) of their children?
   d.   if one of the children marries a person with AKU, what genotypes and phenotypes of children can they have and in what proportions?

Solution
   a.   PKU = *aa BB*,   AKU = *AA bb*
   b.   All *Aa Bb*
   c.   All normal
   d.   *Aa Bb* × *AA bb* -> 1/4 *AA Bb*, 1/4 *AA bb*, 1/4 *Aa Bb*, 1/4 *Aa bb*

11. In *Neurospora*, the *pan-2* (pantothenic acid) locus has the *ylo* (yellow mycelium) locus closely linked to the left, and the *trp-2* (tryptophan) locus closely linked to the right. Three independently isolated *pan-2* alleles, *pan-2a*, *pan-2b*, and *pan-2c*, were used in the crosses listed below. The ascospores from each cross were plated on medium containing tryptophan but no pantothenic acid. The only spores to grow were therefore *pan-2*$^+$ prototrophs. The prototrophs were each tested for the flanking markers and the results are also shown in the table.

Cross	Flanking marker genotypes of *pan-2*$^+$ prototrophs
1 *ylo pan-2a trp* × *ylo*$^+$ *pan-2b trp*$^+$	mostly *ylo trp*$^+$
2 *ylo pan-2a trp* × *ylo*$^+$ *pan-2c trp*$^+$	mostly *ylo trp*$^+$
3 *ylo pan-2c trp* × *ylo*$^+$ *pan-2b trp*$^+$	mostly *ylo*$^+$ *trp*

   a.   Draw diagrams to show the origins of the prototrophs in each cross.
   b.   What is the order of the mutant sites of the *pan-2a, b,* and *c* alleles in relation to the flanking markers?

Solution

a.   Since the protrophs are all recombinant for flanking markers, it is likely that the prototrophs resulted from crossing over between the mutant sites within the *pan-2* gene, and not from reversion.

b.   The order of sites that fit the data is *ylo—c-b-a—trp*.

C H A P T E R

13

DNA Function

Multiple-Choice Questions

1.   In the giant polytene chromosomes of *Drosophila* and other insects, the puffs correspond to
      a.   regions of replication.
      b.   regions of protein synthesis.
      c.   regions of transcription. •
      d.   chromomeres.
      e.   regions of DNA repair.

2.   The following diagram shows a fragment of transcribed DNA, and the upper strand is the template strand:

      5' ATTGCC 3'
      3' TAACGG 5'

The transcribed RNA can be represented by
      a.   5' AUUGCC 3'
      b.   5' TAACGG 3'
      c.   3' AUUGCC 5'
      d.   5' UAACGG 3'
      e.   5' GGCAAU 3' •

3.   Consider the following segment of DNA from within the protein-coding region of a gene

     5′ GGAACTCTAGGGGCTG 3′
     3′ CCTTGAGATCCCCGAC 5′

Which one of the following is true?
   a.   the upper strand must be the transcribed strand
   b.   the lower strand must be the transcribed strand
   c.   either could be transcribed strand •
   d.   both could be transcribed
   e.   neither can be transcribed because both contain stop codons

4.   The following DNA fragment contains the translation initiation codon actually used in the gene.

     CGGAACATCGC
     GCCTTGTAGCG

The template strand must be
   a.   upper strand, 5′->3′ as written. •
   b.   upper strand, 3′->5′ as written.
   c.   lower strand, 5′->3′ as written.
   d.   lower strand, 3′->5′ as written.
   e.   several possibilities.

5.   In a chromosome one of the following is true.
   a.   The RNAs of all genes are synthesized 5′->3′ off the same DNA strand.
   b.   The RNAs of all genes are synthesized 3′->5′ off the same DNA strand.
   c.   RNAs of different genes can be transcribed off either DNA strand, but always 5′->3′. •
   d.   RNAs of different genes can be transcribed off either DNA strand, but always 3′->5′.
   e.   Different genes can be transcribed off either strand, some in 5′->3′ direction, and some 3′->5′.

6.   The following diagram shows a protein-coding gene and specifies the positions of its translational start and stop signals

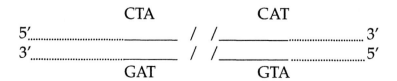

Which of the following is correct?
   a.   upper is template strand, carboxy terminus of protein to right
   b.   upper is template strand, carboxy terminus of protein to left •
   c.   lower is template strand, carboxy terminus of protein to right
   d.   lower is template strand, carboxy terminus of protein to left
   e.   either strand could be template based on this information

7.   A protein is 300 amino acids long. In the coding region for that gene, the total number of nucleotide pairs is
   a.   100.
   b.   300.
   c.   600.
   d.   900. •
   e.   1200.

8.   An amber mutation (UAG) in yeast is crossed to an amber suppressor. In the resulting ascospore population, what percentage will be wild-type?
   a.   0
   b.   25
   c.   50
   d.   75 •
   e.   100

9.   In a haploid eukaryote, an auxotrophic mutant whose mutational site is an amber mutation (UAG) is crossed to a strain carrying two separate independently segregating amber suppressors. What proportion of ascospores will be auxotrophic?
   a.   1/8 •
   b.   1/4
   c.   1/2
   d.   3/4
   e.   7/8

10.   (Codon dictionary needed.) A protein fragment has the following sequence
      amino end - *his-met-leu-ile-lys* - carboxyl end

The DNA template strand sequence (3'-5') would be
   a.   GUGUACAAUAUUUUU.
   b.   GTGTACAATATTTTT.
   c.   GTATACAATTATTTT. •
   d.   GUAUACAAUUAUUUU.
   e.   CATATGTTAATAAAA.

11. Which of the following is found in proteins but not DNA?
    a. phosphorus
    b. sulfur •
    c. oxygen
    d. hydrogen
    e. nitrogen

12. The association of separate amino acid chains to constitute one active protein is called the
    a. primary structure.
    b. secondary structure.
    c. tertiary structure.
    d. quaternary structure. •
    e. quinternary structure.

13. (Coding dictionary required) A tRNA with the anticodon 3' ACC 5' would carry
    a. phenylalanine.
    b. tyrosine.
    c. tryptophane. •
    d. serine.
    e. threonine.

14. Which of the following is not a component of the translational system in cells?
    a. ribosome
    b. RNA polymerase •
    c. tRNA
    d. aminoacyl-tRNA transferase
    e. mRNA

15. All of the following components except one are general purpose translation components which could in principle be used in the translation of any gene.
    a. tRNA
    b. ribosome
    c. phenylalanine
    d. mRNA •
    e. initiation factors

16. In a wild-type strain of *Drosophila* the size of a gene from start codon to stop codon is calculated to be 2000 nucleotide pairs. However, the size of the mRNA molecule transcribed from this gene is consistently estimated to be 1200 nucleotide pairs. The most likely explanation is the existence of
    a. a new stop codon introduced by mutation.
    b. a frame shift mutation.
    c. mRNA degredation.
    d. DNA degredation.
    e. presence of one or more introns. •

17. Which of the following acts before the others
    a. tRNA alignment with mRNA
    b. RNA polymerase •
    c. aminoacyl-RNA synthetase
    d. ribosome movement to next codon
    e. amino acid chain elongation

Open-Ended Questions

1. Consider the following piece of messenger RNA

5' AUGGAGUCGUUAAUUAAACCGGTGCGGATCGTAUUUAGUCCCCAC 3'

   a. Draw both strands of the segment of DNA that this mRNA was transcribed from.
   b. State which of the DNA strands the RNA synthesizing enzyme used as template for the transcription process.
   c. Show the amino acid sequence of the protein that would be produced by translation of the mRNA, assuming that the ribosome moved along the mRNA from left to right. (Use a code dictionary.)

Solution
   a. 3' TACCTC............etc......................GGGGTG 5'
      5' ATGGAG.........etc......................CCCCAC 3"

   b. The top strand (3' – 5')
   c. *met.glu.ser*.......etc...............*ser.pro.his*

2. (Codon table required) A gene in *E. coli* codes for an enzyme with 100 amino acids; it has the following amino acid sequence near one end

   -trp-val-pro-leu-phe-asp-glu-arg-

A mutant strain lacks activity for this enzyme and the enzyme has the following sequence in the same region (all other amino acids are identical)

   -trp-val-his-tyr-ser-asn-glu-arg-

   a. Deduce the mRNA nucleotide sequence from which these proteins were synthesized.
   b. State the nature of the mutational change.
   c. Since the mutant's enzyme was very similar to the wild-type, why was the strain mutant?

Solution

    a.   +   UGG GUN CCA UUA UUC GAU/C GAA/G CGN
          mut UGG GUN CAU UAU UCG AAU/C GAA/G CGN

    b.   Deletion of a GC pair in the DNA and an inserion of a TA pair.
    c.   The four different amino acids probably were at an active site for the enzyme and resulted in an inability to bind or interact with substrate molecules.

3.   In corn, synthesis of purple pigment is controlled by two genes acting sequentially through colorless ('white') intermediates

    white 1————————————>white 2————————————>purple
               gene A   gene B

UAG nonsense mutations were obtained in the genes A and B, and we can call them a^n and b^n. These both lacked enzyme activity and were recessive to their wild-type alleles A and B. A nonsense mutation is caused by a nucleotide change that results in a stop codon; in this case the same nonsense codon was produced, UAG. A nonsense suppressor is also available; it is a mutation in a duplicate copy of a tRNA gene. The mutational site is in the anticodon and this causes the tRNA to insert the amino acid tryptophan at the nonsense codon UAG, thereby allowing the protein synthesis machnery to complete translation, and the result is a wild-type phenotype. We will represent the nonsense suppressor allele as T^s and its wild-type allele T^+. Neither T^s nor T^+ have any detectable effect on the wild-type alleles A and B or on any other aspect of the phenotype.

    a.   Would you expect T^s to be dominant to T^+ or not? Explain.
    b.   A trihybrid $Aa^n\ Bb^n\ T^sT^+$ is selfed. If all the genes are unlinked, what phenotypic ratio do you expect in the progeny. Explain carefully. (Assume absence of purple results in a light color that we can call white.)

Solution

    a.   Dominant because the suppressor allele is active in the system, whereas the wild-type allele is not.

    b.   9/16   purple   A- B-   36/64 purple
          7/16   white   A- bb $\times$ 3/4 T^S- -> 21/64 purple
                               aa B- $\times$ 1/4 T^+T^+ -> 7/64 white
                               aa bb
    Totals: 57/64 purple and 7/64 white.

4.   In a haploid organism, a short stretch of amino acid sequence was obtained for a certain protein P from several different mutants with mutational sites in that region. The amino acid sequences were as follows

Wild-type	*met-leu-lys-arg-glu*
Mutant 1	*met-leu* (no further amino acids from here on)
Mutant 2	*met-leu-glu-thr-arg*-etc (all different from here on)
Mutant 3	*met-leu-lys-cys-glu*

a.   Use the codon table to derive a possible nucleotide sequence for this region and deduce the molecular nature of the mutations.

b.   If mutants 1 and 2 are intercrossed, and 100 progeny isolated, approximately how many are expected to be wild-type? Explain.

Solution

a.   In mutant 1 a *lys* codon AAG mutated to UAG.

b.   Insertion of a G caused frame shift

> *lys*
> AAACGCGAA...
> GAAACGCGA...
> *glu*

c.   *arg* to *cys* was a missense mutation for example, CGU to UGU

5.   From DNA sequencing studies the nucleotide sequences of three short adjacent fragments (F_1, F_2, and F_3) of a mouse gene are as follows, grouped as codons

F_1 3′ AGA.GCC.ATG.TTT.CCT 5′
F_2 3′ CCT.TAC.ACA.CCA.GAA 5′
F_3 3′ ACA.CCA.ACT.CCT.TTT 5′

a.   Write out the RNA sequence coded by each frgment.

b.   Order the three fragments and write out the amino acid sequence.

c.   Mark the amino and carboxy termini of the protein as written.

Solution

a.   F1 5′ UCU CGG UAC AAA GGA 3′
F2 5′ GGA AUG UGU GGU CUU 3′
F3 UGU GGU UGA GGA AAA 3′

b.   There is a single AUG initiation codon (in F_2) and a terminator in F_3 so order is F_2 - F_1 - F_3 and the amino acid sequence must be

met cys gly leu ser arg tyr lys gly cys gly

c.   amino on left, carboxy on right.

6.   Consider the following segment of the DNA of the protein-coding region of a gene. It is composed of 18 nucleotide pairs and these represent 6 codons. One is shown in bold (see following page).

Template strand ->   3′ CTT- GGC-**GTT**- TTC-GGA-GTA 5′
                                    5′ GAA-CCG-**CAA**-AAG-CCT-CAT 3′

    a.   Illustrate the process of replication using this sequence and draw the two daughter molecules that result (label with 5′, 3′).
    b.   Draw the mRNA that is transcribed from this sequence (label with 5′, 3′).
    c.   Draw the amino acid chain that is encoded by this sequence.
    d.   Draw the anticodon of the tRNA that aligns with the bold codon (label with 5′, 3′).

Solution
    a.   Both daughter molecules will be identical to original. Note 5′->3 synthesis will be leading strand, 3′->5′ will be lagging strand.
    b.   mRNA will be 5′ GAA CCG CAA AAG CCU CAU 3′
    c.   *glu-pro-gln-lys-pro-his*
    d.   3′-GUU-5′

C H A P T E R

14

Recombinant DNA Technology

Multiple-Choice Questions

1. In recombinant DNA technology, DNA is most often cut using
    a. DNA ligase.
    b. DNA polymerase.
    c. DNA gyrase.
    d. restriction endonucleases. •
    e. terminal transferase.

2. Which of the following has not been used as a cloning vector?
    a. F factor
    b. resistance plasmid
    c. cosmid
    d. virus
    e. autonomously replicating introns •

3. A circular DNA molecule has three target sites for restriction enzyme *Eco*RI. How many fragments will be produced after complete digestion?
    a. 1
    b. 2
    c. 3 •
    d. 4
    e. 5

4.   A linear DNA molecule has three target sites for restriction enzyme *Eco*RI. How many fragments will be produced after complete digestion?
    a.   1
    b.   2
    c.   3
    d.   4•
    e.   5

5.   A linear DNA molecule has three target sites for restriction enzyme *Eco*RI. What is the maximum number of fragments that can be produced if a sample of the molecule is only partly digested?
    a.   10 •
    b.   3
    c.   5
    d.   4
    e.   7

6.   A plasmid vector is cut in one place and a poly G tail is added to the 3′ ends. For effective cloning, the donor DNA should have
    a.   poly G at the 3′ ends.
    b.   poly G at the 5′ ends.
    c.   poly C at the 3′ ends. •
    d.   poly C at the 5′ ends.
    e.   poly G at the 5′ ends and poly C at the 3′ ends.

7.   A plasmid vector has a gene for erythromycin resistance and a gene for ampicillin resistance. The *amp* gene is cut with restriction enzyme, and donor DNA treated with the same enzyme is added. What genotype of cells needs to be selected after transformation?
    a. str^R
    b. $amp^R\ ery^R$
    c. $amp^R\ ery^S$ •
    d. $amp^S\ ery^R$
    e. $amp^S\ ery^S$

8.   Assume a cosmid will carry inserts of about 50 kb and cosmids are used to clone a genome of size 3 Megabases (1 Megabase = 1 million bases). Assuming the best possible luck, what is the smallest number of cosmid clones that could constitute a genomic library?
    a.   60 •
    b.   300
    c.   30
    d.   3000
    e.   16.66

9.   A clone of the actin gene from yeast is used to probe an *Eco*RI digest of *Neurospora* genomic DNA using a Southern analysis. The autoradiogram shows a single-labelled band of 6 kb in size. This means
  a.   the *Neurospora* gene is 6 kb in size.
  b.   the *Neurospora* gene could be greater than 6 kb in size.
  c.   the *Neurospora* gene does not have any introns.
  d.   *Neurospora* just has one copy of the actin gene. •
  e.   the *Neurospora* gene is the same size as the yeast gene.

10.   The cDNA for a eukaryotic gene B is 900 nucleotide pairs long. A cDNA clone is used to isolate a genome clone of gene B and the gene is sequenced. From start to stop codon the gene is found to be 1800 nucleotide pairs long. This discrepancy is probably because of
  a.   the mRNA broke during cDNA synthesis.
  b.   the gene is a present as a tandem duplicate.
  c.   the gene has 900 np of introns. •
  d.   the genomic clone is not really gene B, just a related gene.
  e.   a sequencing error.

11.   A clone from a genomic library is used to transform a purine-requiring mutant (*pur-3*) to prototrophy. The transformant is test crossed to a *pur-3* strain of opposite mating type and 1/2 of the progeny are purine-requiring. This probably means
  a.   the wild allele inserted at the original *pur-3* locus.
  b.   the wild allele inserted ectopically •
  c.   the tester parent could not have been mutant for the same pur locus.
  d.   a new *pur* mutation arose.
  e.   the transformed strain was a mixture of transformed and untransformed nuclei.

12.   A wild-type *Aspergillus* strain is transformed with a plasmid carrying a hygromycin resistance allele and cells are plated on hygromycin. One resistant colony showed an aberrant type aerial hyphae. When crossed to wild-type the progeny were 1/2 normal hyphae *hyg* sensitive, and1/2 aberrant hyphae *hyg* resistant. The probable explanation is
  a.   the plasmid inserted in a gene for normal hyphal development. •
  b.   the plasmid interfered with hyphal development.
  c.   a mutation arose in a gene for hyphal development on the plasmid.
  d.   a mutation arose in a gene for hyphal development on a recipient chromosome.
  e.   the recipient was a heterokaryon carrying some mutant nuclei.

13.   In the Sanger sequencing method the use of dideoxy adenosine triphosphate stops nucleotide polymerization
    a.   opposite A's in the template strand.
    b.   opposite T's in the template strand. •
    c.   opposite G's in the template strand.
    d.   opposite C's in the template strand.
    e.   opposite any base selected randomly in the template strand.

14.   In any fragment of double-stranded DNA how many possible different reading frames are there?
    a.   2
    b.   3
    c.   4
    d.   5
    e.   6 •

15.   A 1 kb gene is labelled at one end with ^{32}P and partially digested with restriction enzyme *Bgl*II. Electrophoresis followed by Southern blotting and autoradiography reveals four bands of sizes 0.1, 0.4, 0.6, and 1 kb in size. The *Bgl*II restriction sites are separated by the following distances in the order shown (starting from the labelled end).
    a.   0.1-0.4-0.6-1.0
    b.   1.0-0.6-0.4-0.1
    c.   0.4-0.2-0.3-0.1
    d.   0.1-0.3-0.2-0.4 •
    e.   1.0-1.1-1.4-1.6

Open-Ended Questions

1.   You have isolated a globin gene from a hamster genomic library. The DNA is 15 kb long. You digest it with the restriction enzymes *Eco*RI and *Sal*I. The results are as follows

EcoRI alone: three fragments of 3 kb, 4 kb, and 8 kb
*Sal*I alone: four fragments of 1 kb, 3.5 kb, 5 kb, and 5.5 kb
EcoRI cuts: the 5.5 kb *Sal* fragment into 2.5 kb and 3 kb.
                the 3.5 kb *Sal* fragment into 1.5 kb and 2 kb
*Sal*I cuts: the 4 kb *Eco* fragment into 2.5 kb and 1.5 kb
                the 8 kb *Eco* fragment into 1 kb, 2 kb and 5 kb.
    a.   Draw a restriction map of this globin gene
    b.   If a double digest is electrophoresed, and a Southern blot probed with labeled
    c.   DNA from the same gene, the label binds only to the 1.5 kb, 3 kb, and 5 kb fragments. Provide an explanation for this hybridization pattern.

Solution

    a.   end-3-E-2.5-S-1.5-E-2-S-5-S-1-end

    or   S-1-S-5-end

    b.   There must be introns in the gene.

2.   The chromosomal region containing a human hemoglobin gene (the gene that is defective in the autosomal recessive disease sickle cell anemia) contains three sites for the restriction enzyme *Dde*I in the following approximate positions in the normal sequence (the square brackets show the extent of the gene).

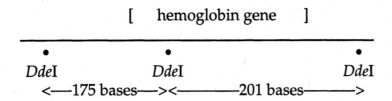

    [hemoglobin gene   ]

    *Dde*I   *Dde*I   *Dde*I

    <—175 bases—><————201 bases————>

*Dde*I cuts DNA at the target sequence CTNAG where N is any base. Near one end of the normal hemoglobin there is the following group of amino acids, one of which is substituted in people with sickle cell anemia

    — pro — glu — glu —

    —CCT —GAG —GAG

whereas the sickle cell hemoglobin has the following amino acids in this position

    — pro — val — glu —

    —CCT —GTG —GAG —

    a.   If you had a clone of this hemoglobin gene, how could you use the *Dde*I enzyme to distinguish the DNA of normal people from that of sickle cell anemia patients?

    b.   How could you use the *Dde*I enzyme to identify people who are heterozygous for the sickle cell allele?

Solution

    a.   The mutation that causes the amino acid substitution destroys a *Dde*I site, which must be the one shown in the hemoglobin gene. Therefore, in a Southern blot of a *Dde*I genomic digest probed with the hemoglobin gene there should be two radioactive bands on the autoradiogram, of 175 and 210 bp in size. However, people who are homozygous for sickle cell anemia should show only a single large fragment of 376 bp.

    b.   heterozygotes should show three bands, of 175, 201 and 376 bp in size.

3.   A 32 kb circular plasmid was digested with the following restriction enzymes: *Eco*RI (E), *Pst*I (P), and *Kpn*I (K). Gels of the digests are shown on the following page, fragment lengths are in kb.

E	P	K
	_ 32.0*	
_ 26.0*		
		_ 19.0
		_ 13.0*
_ 6.0		

E & K	E & P	P & K
	_ 20.5	
		_ 19.0
_ 13.0*		
		_ 12.0
_ 8.5		
_ 6.0	_ 6.0	
	_ 5.5*	
_ 4.5		
		_ 1.0*

a.   From the above data draw a complete restriction enzyme map of this plasmid showing all the restriction sites and fragment lengths between sites.

b.   The plasmid in part a contains an insert between two of the RE sites shown. When the insert was used as a probe in a Southern blot of the gel from part a the bands indicated with an asterisk (*) showed up on the autoradiogram. Indicate clearly on the plasmid where the insert is located.

Solution
    a. and b.

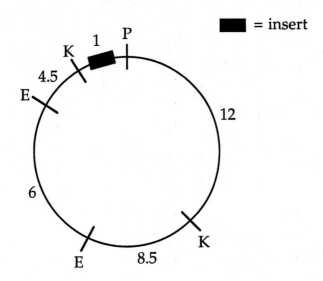

4.   A 5 kb fragment was 5′ end-labelled at both ends with ^{32}P. It was then divided into two aliquots and one was digested with *Bam*HI, the other with *Eco*RI. Each cut at only one site in the DNA. The largest fragment from each digest was purified and then partially digested with *Sau*3A (a frequent cutter).

    After running each partial digest on a gel and exposure to film, the result below was obtained. From this data draw a restriction map of the original 5 kb fragment showing *Bam*, *Eco*, and *Sau* sites.

*Bam*HI + *Sau*3A	*Eco*RI + *Sau*3A
—— 4.6	
—— 4.4	—— 4.4
	—— 3.9
—— 2.6	
	—— 2.4
—— 1.1	
	—— 0.6
—— 0.5	

Solution
0.5-S-0.1-E-0.5-S-1.5-S1.8-S-0.2-B-0.4

5.   A Sanger dideoxy sequence analysis was performed on a region of cloned wild-type DNA and the same region in a mutant strain. The sequencing gels are represented below.

From these results
    a.   determine each sequence in full.
    b.   determine the mutant site and the nature of the mutational event.

Solution
    a.   CCAAT(C/T)TTGTGCAC
    b.   A GC nucleotide pair in the wild-type had changed to an AT pair (C->T on the sequenced strand)

6.   A 10 kb linear DNA molecule was digested with restriction enzyme *Eco*R1 and two fragments were produced of sizes 6 kb and 4 kb. The restriction enzyme *Hae*III also produced two fragments but of sizes 9 kb and 1 kb. When the two restriction enzymes were used together three fragments were produced of sizes 1 kb, 3 kb, and 6 kb. A cloned probe from this region hybridized to the 1 kb and 6 kb segments in a Southern analysis. Draw a map of the arrangement of the restriction enzyme target sites on this molecule.

Solution
end-6kb-E-1 kb-H-3kb-end

CHAPTER

15

Applications of Recombinant DNA Technology

Multiple-Choice Questions

1.   A cloned yeast gene of unknown function was subjected to in vitro mutagenesis and a serine codon was replaced by arginine at amino acid position 10 in the open reading frame (ORF). This gene was used to replace the resident wild-type gene. The resulting cell still showed wild-type phenotype. The probable reason is
   a.   the amino acids have equivalent function at that position.
   b.   there is another copy of the wild-type gene present in the genome.
   c.   the mutant gene did not replace wild-type but inserted ectopically.
   d.   the gene has no function; it is an inactive 'pseudogene'.
   e.   any of the above could be true. •

2.   In a gene disruption experiment, the recipient cells must
   a.   be wild-type for the gene in question. •
   b.   be mutant for the gene in question.
   c.   be deleted for the gene in question.
   d.   have a drug resistance marker in the gene in question.
   e.   be of opposite mating type.

3.   A fertilized egg from a dwarf mouse line (caused by a homozygous recessive mutation *lit lit*) is injected with a normal active rat growth hormone gene and the resulting transgenic mouse is normal-sized. This transgenic mouse is crossed to a *lit lit* mouse. What proportion of progeny will be normal-sized if there was only one transgene?
>   a.   0
>   b.   1/4
>   c.   1/2 •
>   d.   3/4
>   e.   1/3.

4.   Two fertilized eggs from a dwarf mouse line (caused by a homozygous recessive mutation *lit lit*) were injected with a normal active rat growth hormone gene and the resulting transgenic mice were normal-sized. The two transgenic mice happened to be one male and one female, and these were crossed. What proportion of progeny will be normal-sized if there was only one transgene?
>   a.   0
>   b.   1/4
>   c.   1/2
>   d.   3/4 •
>   e.   1/3.

5.   A certain probe detects an *EcoR* I-based RFLP in *Neurospora*: one morph shows a fragment of 1 kb and the other morph a fragment of 2 kb. If these two morphs are crossed the resulting ascospores will
>   a.   show no RFLP morphs.
>   b.   be 50% the 1 kb morph and 50% the 2 kb morph. •
>   c.   all show a 3 kb morph.
>   d.   all show a 1.5 kb morph.
>   e.   be 50% a 3 kb morph and 50% a 1.5 kb morph.

6.   A man is heterozygous for a disease gene D and also for an RFLP linked at a distance of 4 map units. The chromosomal arrangement is
D morph-1 / d morph-2. The percent of his sperm that will be D morph-2 is
>   a.   50%
>   b.   25%
>   c.   4%
>   d.   2% •
>   e.   46%

7.   A certain strain of a haploid fungus is transgenic for a firefly gene that is not normally found in this fungus. The transgenic strain is crossed to a normal strain. PCR primers based on the firefly sequence are used in a PCR analysis of individual ascospore progeny. What percentage of ascospores will show amplifications?
    a.   0%
    b.   25%
    c.   50% •
    d.   75%
    e.   100%

8.   A certain strain of a haploid fungus is transgenic for a firefly gene that is not normally found in this fungus. The transgenic strain is crossed to a normal strain. PCR primers based on the sequence of the resident fungal gene for actin protein are used in a PCR analysis of individual ascospore progeny. What percentage of ascospores will show amplifications?
    a.   0%
    b.   25%
    c.   50%
    d.   75%
    e.   100% •

9.   A woman is heterozygous for a dominant disease gene H, and also for an RFLP linked 10 map units away. The chromosomal arrangement is H RFLP morph-1 / h RFLP morph-2. Her unaffected husband is homozygous for morph-2. A fetus is tested and is shown to be homozygous for morph-2. What is the probability that, if born, the child will have the disease?
    a.   10% •
    b.   90%
    c.   5%
    d.   45%
    e.   20%

10.   The mutation causing the recessive disease allele for sickle cell anemia removes one *Mst*II restriction site from the globin gene so that a probe instead of hybridizing to one 1.3 kb fragment hybridizes to two fragments of 1.1 kb and 0.2 kb. For two parents to have a 25% chance of a child with sickle cell anemia
    a.   they must both show only fragments of 1.1 and 0.2 kb.
    b.   they must both show fragments of 1.3 and 0.2 kb.
    c.   one must show only a 1.1 and the other only a 0.2 kb fragment.
    d.   one must show a 1.3 and the other a 1.1 and a 0.2 kb fragment.
    e.   they must both show all three types of fragments. •

11.   A wild-type mouse blastocyst is injected with cells heterozygous for a dominant mutation for bent tail. If this blastocyst develops into a mouse and the mouse is crossed to wild-type, what percent of progeny will have a bent tail?
    a.   0%
    b.   25%
    c.   50%
    d.   100%
    e.   can't tell •

12.   A transgenic *Arabidopsis* plant contains two copies of the transgene T, one on chromosome 1 and one on chromosome 2. What percent of gametes will not contain the transgene?
    a.   2%
    b.   10%
    c.   25% •
    d.   50%
    e.   100%

13.   A *ura*⁻ yeast cell is transformed to *ura*⁺ by a cloned fragment that inserts ectopically. If this transformant is crossed to wild-type then the proportion of *ura*⁻ progeny is
    a.   0%
    b.   25% •
    c.   50%
    d.   75%
    e.   100%

14.   A *ura*⁻ yeast cell is transformed to *ura*⁺ by a cloned fragment that replaces the endogenous gene. If this transformant is crossed to wild-type then the proportion of *ura*⁻ progeny is
    a.   0%
    b.   25%
    c.   50%
    d.   75%
    e.   100% •

15.   In a population of the diploid plant *Mimulus guttatus*, a probe detects a total of five different RFLP morphs. In any one plant the maximum number of RFLP morphs that could be detected is
    a.   1.
    b.   2. •
    c.   3.
    d.   4.
    e.   5.

Open-Ended Questions

1.   Children with Williams syndrome are small, mentally retarded and have highly characteristic 'elfin' features. It is most often sporadic but can be inherited as an autosomal dominant disease. In 90% of cases, a cloned gene for the protein elastin hybridizes in situ to only one homolog on the long arm of chromosome 7. In the remainder of cases the probe binds to both homologs.

    a.   Provide a genetic explanation for these two types of Williams syndrome.

    b.   Why do you think the disease is always inherited as a dominant and not a recessive?

Solution

    a.   First type is most likely a microdeletion surrounding the elastin gene: the deletion may or may not be visible microscopically. The second type is most likely a point mutation in the elastin gene, causing null or reduced function.

    b.   Probably because the gene is haplo-insufficient, in other words in the heterozygote a single wild-type copy cannot provide enough function for normal cell operation.

2.   The restriction enzyme Sau_I site S is known to be closely linked to the left side of the gene that causes Huntington's disease, a severe neurological disease. There is a restriction enzyme polymorphism (RFLP) involving the S site and other sites to the left of S. There are three common alleles, of 1, 2, and 3 kilobases in size as shown in the following diagram

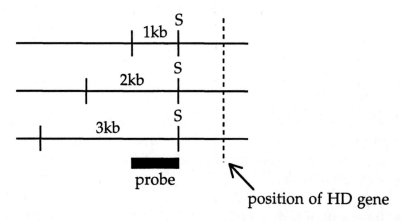

position of HD gene

Consider the following pedigree for HD.

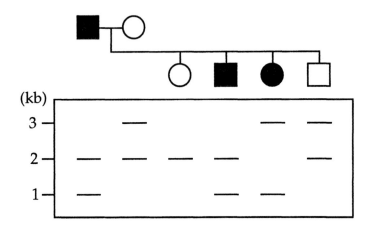

The fourth child (a son) has just been born and the parents want to know if he will develope HD (which you will remember is a late onset disease).

    a.   You are called in as a genetic consultant. What is your opinion?

    b.   Draw a diagram that explains your answer to the parents.

Solution

    a.   The 1 kb allele from the father must be the one linked to the HD allele. The last son does not have this allele so he will most likely not develop HD.

    b.

```
_____S___H

_____S_____S___h

_S_____S___h
```

3.   Assume that a certain piece of human DNA of 1 kilobase (kb) in size, when cloned and used as a probe, reveals an RFLP in total DNA cut with the restriction enzyme EcoRI. Individuals can show one of the following patterns

    i.   one fragment of 6 kb

    ii.   two fragments of 4 kb and 2 kb

    iii.   three fragment of 6 kb, 4 kb and 2 kb.

    a.   Draw diagrams that show you understand what the above statements mean.

    b.   In a certain family showing individuals with a late onset autosomal dominant disease, it was found found that all the people with the disease showed pattern (iii) and unaffected individuals were all of pattern (ii). How do geneticists explain this kind of association?

    c.   If a man in the family is recently married and contemplating having a child, what tests need to be done?

Solution

a.

E_____6 kb_____E haplotype 1

E____2 kb____E_____4 kb_____E haplotype 2

XXXXX
probe

Individuals can be 1/1, 2/2, or 1/2.

b.   In this family the disease allele D is linked to the type 1

D_____E_____6_____E
d_____E_2__E_____4_____E

c.   The man should have a DNA test to determine if he carries D (look for type 1). If so he might choose not to have children.

4.   In a certain community a widespread mutant allele for Tay-Sachs disease (autosomal recessive) removes a *Hind*III restriction site in the gene sequence. A probe is available for this region, with the following homology:

Tay-Sachs mutant region __H_____H____
Normal sequence   _H_____H_____H____
Probe homology   _____

Two couples in this community are expecting a child and they have DNA tests to determine if the child will have Tay-Sachs.

Autoradiogram

Mother 1	Father 1	Fetus	Mother 2	Father 2	Fetus 2
___	___	___	___	___	___
___	___		___	___	___

How would you counsel the couples?

Solution
Both parents in couple 1 are heterozygous for the normal and mutant allele so 1/4 of their children will have Tay-Sachs disease. The particular fetus in question is homozygous for the long fragment indicating that it is homozygous for the Tay-sachs

allele and would develop the disease. Couple 2 are also heterozygous, but their child is also heterozygous and will not express the disease. (These heterozygotes are expected at a frequency of 1/2.)

5.   Probes P and Q detect two linked RFLPs in *Hind*III-digested DNA of *Neurospora crassa*. The two morphs detected by probe P are Morph P1 (fragments of 0.4 and 0.5 kb) and morph P2 (fragments of 0.4 and 0.2 kb). Probe Q detects morph Q1 (0.6 kb) and Q2 (0.8kb). A strain with P1 and Q1 is crossed to a strain with P2 and Q2. Draw the autoradiogram of a tetratype ascus with regards to these markers, hybridized by a mixture of probes P and Q.

Solution

Spore pair	1	2	3	4
0.8 kb		—		—
0.6 kb	—			—
0.5 kb	—		—	
0.4 kb	—	—	—	—
0.2 kb		—		—

6.   An RFLP marker is known to be linked 10 centimorgans away from the locus for brachyrachia (an autosomal dominant spinal defect, allele *B*). In one couple the RFLP probe detected two RFLP 'alleles' *r1* and *r2* in the parent with brachyrachia, and only one allele *r3* in the unaffected parent. The chromosomes were as follows:

$$\frac{B \quad r1}{b \quad r2} \quad \times \quad \frac{b \quad r3}{b \quad r3}$$

The couple is expecting a child and a DNA test is performed on the unborn fetus. What is the probability that the child will have brachyrachia if the DNA test shows the fetal RFLP genotype to be

    a.   *r1 r3*

    b.   *r2 r3*.

Solution

    a.   The gametes will be

        B 1   45%

        B 2   5%

        b 1   5%

        b 2   45%

hence 45/50 (= 90%) of *r1* gametes will have *B* and the disease.

    b.   5/50 (= 10%) of *r2* gametes will have *B*.

C H A P T E R

16

The Structure and Function of Eukaryotic Chromosomes

Multiple-Choice Questions

1. How many DNA molecules are there in a male human cell in metaphase I of meiosis?
   a. 30,000
   b. 60,000
   c. 23
   d. 46
   e. 92 •

2. How many DNA molecules are there in a male human cell in metaphase of mitosis?
   a. 30,000
   b. 60,000
   c. 23
   d. 46
   e. 92 •

3.   How many DNA molecules are there in a male human cell in metaphase II of meiosis?
    a.   30,000
    b.   60,000
    c.   23
    d.   46 •
    e.   92

4.   In a certain fungus the genome size is 4,000 kb. The total chromosome map size is 800 map units. Two loci *A* and *B* show 5% recombinant frequency; how many nucleotide pairs separate them on the chromosome?
    a.   1 kb
    b.   5 kb
    c.   8 kb
    d.   20 kb
    e.   25 kb •

5.   A certain eukaryotic chromosome is 100 kb long. If a restriction enzyme with a six-base target sequence is used to cut this chromosome, approximately how many chromosomal fragments will be produced?
    a.   2
    b.   10
    c.   25 •
    d.   50
    e.   100
(Note: $(1/4)^6$ = 1 cut every 4096 nuc pr, approx. every 4 kb.)

6.   In *Drosophila*, heterochromatin contains
    a.   no genes.
    b.   only nonfunctional genes.
    c.   few genes. •
    d.   the same number of genes per unit of DNA as euchromatin.
    e.   more genes per unit of DNA than euchromatin.

7.   Increasing levels of chromosome packing are
    a.   nucleosomes-solenoids-loops-supercoils. •
    b.   solenoid-nucleosomes-loops-supercoils.
    c.   solenoid-nucleosomes-supercoils-loops.
    d.   nucleosomes-loops-solenoid-supercoils.
    e.   solenoids-loops-nucleosomes-supercoils.

8.   If gene order is a^+-b^+-c^+-hetrochromatin, and all three genes can be subject to position effect variegation, a phenotype that could not be produced is
   a. $a^-b^-c^-$
   b. $a^+b^-c^-$
   c. $a^+b^-c^-$
   d. $a^+\,b^+c^+$
   e. $a^-b^+c^+$ •

9.   Consider a +/m hetrozygote in which the locus is next to heterochromatin. In comparison to wild-type, a *Su*(var) strain would show
   a.   more tissue expressing m.
   b.   less tissue expressing m. •
   c.   finer grain mosaicism for + and m.
   d.   coarser grain mosaicism for + and m.
   e.   no expression of + at all.

10.   Which type of DNA segment does code for a protein?
   a.   centromere
   b.   telomere
   c.   rDNA
   d.   nucleolus organizer
   e.   ORF •

11.   A cell yeast normally has 22 chromosomes; if it also contains a YAC, how many centromeres are there in the cell in G1 phase of the cell cycle?
   a.   18
   b.   22
   c.   23 •
   d.   44
   e.   46

12.   How many telomeres are there in a human cell in mitotic metaphase?
   a.   1
   b.   2
   c.   23
   d.   46
   e.   92 •

13.   Which of the following is not a type of repetitive DNA?
   a.   satellite DNA
   b.   VNTRs
   c.   SINEs
   d.   LINEs
   e.   Giemsa band •

14. DNA fingerprints are based on
    a. VNTRs •
    b. SINES
    c. LINES
    d. *Alu* sequences
    e. Giemsa bands

15. The replication problem at telomeres is
    a. DNA polymerase can't attach.
    b. spindle fibers can't attach.
    c. leading strand has no template.
    d. can't prime lagging strand. •
    e. chromatin too tightly coiled.

Open-Ended Questions

1. A clone of a functional gene coding for a subunit of ATPase of mice was obtained from a cDNA library. Human genomic DNA was digested with a six-base target sequence restriction enzyme, and the mouse cDNA was used as a probe in a Southern analysis. The autoradiogram showed three radioactive bands. What explanations can you think of? (List two.)

Solution
    a. If there is only one functional locus for this gene and the enzyme does not cut within it, then only one band is expected. If the enzyme cuts the gene then more than one band will be seen.
    b. Alternatively there might be several nonfunctional pseudogenes in the genome that will hybridize the probe.

2. An enhancer of position-effect variegation in *Drosophila* produced variegation in genes (a, b, and c) that were not previously subject to position-effect variegation, and whose order was not known. A fly of genotype a⁺b⁺c⁺/a⁻b⁻c⁻ showed the following mutant sectors

c⁻ (common)

a⁻ c⁻ (less common)

a⁻b⁻c⁻ (rare)
    a. What was the order of the genes in relation to the heterochromatin, and
    b. why were some sectors more common than others?
    c. Why were there no other types of mutant sectors?
    d. Why did these loci not normally show position effect variegation?

Solution

    a.   The order is b-a-c-heterochromatin

    b.   c was nearest so was engulfed most often

    c.   It would be impossible (for example) to have a b⁻ sector because the heterochromatin engulfing process would have to 'jump over' the *a* and *c* loci.

    d.   The mobile edge of the heterochromatin normally did not extend this far but did in the E(var) strain.

3.   The hereditary disease Marie-Chilcote Tooth syndrome has been found to be caused by a tandem duplication of a short segment of about 30 genes. Molecular studies have shown that at the outer ends of the duplicated region, and in the middle between duplicates, there is apiece of DNA resembling a *Drosophila transposon*. Using this information propose a mechanism for the origin of this disease.

Solution

If the transposon (Tn) is present as a type of repetitive DNA, and specifically is found at each end of the segment in normal chromosomes (that is, Tn-30 genes-Tn), then it could act as a substrate for an unequal crossover,

      Tn-30 genes-Tn
             X
      Tn-30 genes-Tn

one product of which would be of the constitution Tn-30 genes-Tn-30 genes-Tn.

4.   A mouse with a chinchilla-colored coat (caused by a recessive autosomal mutation (*c*) was crossed to a mouse of wild-type phenotype but was heterozygous for an inversion of one long chromosome arm. Some of the progeny showed chinchilla spots on a wild-type background, but others were wild-type in appearance. Propose a genetic explanation for both these offspring.

Solution

Those mice who inherited the inversion showed position-effect variegation because the wild-type *c* allele must have been on the inverted segment and is now next to heterochromatin. The wild-type progeny inherited the normal chromosome so the *c* allele is not expressed as expected in normal heterozygotes.

5.   A cloned fragment of mouse DNA of unknown origin was used as a probe in an in situ hybridization experiment. The probe hybridized along the length of each chromosome. What type of DNA could this have been?

Solution

Some type of highly repetitive dispersed sequence; from the information given it is not possible to tell which one.

C H A P T E R

17

Genomics

Multiple-Choice Questions

1. For making libraries of large genomes such as those of mammmmals, the best vector is
   a. YAC. •
   b. pUC18.
   c. pBR322.
   d. lambda phage.
   e. cosmid.

2. The best evidence to prove that a candidate gene is a disease gene is
   a. finding a start and stop codon.
   b. finding a CpG island upstream.
   c. finding the gene is expressed.
   d. finding the homologous gene in a lot of similar animals.
   e. finding a mutation in the homologous sequence from a disease sufferer. •

3. In a certain diploid plant $2n = 24$ and all the chromosomes are small. If DNA is extracted from plant tissue in mitotic metaphase, and run on a pulsed-field electrophoresis gel, how many DNA bands should be visible?
   a. 6
   b. 12 •
   c. 24
   d. 48
   e. 96

4.   In a certain diploid plant $2n = 24$ and all the chromosomes are small. If DNA is extracted from plant tissue in meiotic prophase I, and run on a pulsed-field electrophoresis gel, how many DNA bands should be visible?

    a.   6
    b.   12 •
    c.   24
    d.   48
    e.   96

5.   In the haploid fungus *Neurospora* $n = 7$. A strain bearing a reciprocal translocation is subjected to pulsed-field gel electrophoresis. How many DNA bands are expected to be seen on the gel?

    a.   14
    b.   6
    c.   7 •
    d.   8
    e.   15

6   Four clones (A, B, C, and D) of human genomic DNA are tested for sequence-tagged sites 1 through 5. A shows 2 and 3; B shows 2 and 5; C shows 1 and 5; D shows 3 and 4. What is the order of the clones in their contig?

    a.   ABCD
    b.   BADC
    c.   CBAD •
    d.   ACBD
    e.   DCAB

7.   In genome mapping the general procedure uses the methods RFLP mapping (R), FISH (F), DNA sequencing (S) and physical mapping (P) in the order

    a.   RFPS.
    b.   PSRF.
    c.   SRFP.
    d.   FRPS. •
    e.   PRFS.

8.   In a FISH analysis, a cloned probe of a single-copy gene is hybrid to cells in mitotic metaphase. How many flourescent spots will be seen in one cell?

    a.   1
    b.   2
    c.   3
    d.   4 •
    e.   too many to count

9.   In a human/rodent hybrid cell analysis, 20 clones are examined; none of these have human chromosome 10 and none have the human enzyme galactokinase. The gene for galactokinase
   a.   must be on chromosome 10.
   b.   cannot be on chromosome 10.
   c.   might be on chromosome 10. •
   d.   must be on a chromosome other than 10.
   e.   cannot be on a chromosome other than 10.

10.   From a *Neurospora* cross

   RFLP-1^O *ad-2* x RFLP-1^M *ad-2$^+$*

ad-2$^+$ progeny are isolated and 85% of them are found to carry

RFLP-1^M. (O is an RFLP locus 1 allele from Oak Ridge and M is an allele from Mauriceville). The map distance between the RFLP-1 locus and the *ad-2* locus is
   a.   85 cM.
   b.   30 cM.
   c.   15 cM. •
   d.   7.5 cM.
   e.   42.5 cM.

11.   In an irradiation and gene transfer experiment assume that human gene P is transferred to 6% of hybrids and gene Q to only 2%. Hybrids with both P and Q were found at a frequency of 0.12%. This result shows
   a.   P and Q are linked quite close together on the same chromosome.
   b.   P and Q must be on different chromosomes.
   c.   P and Q are at opposite ends of the same chromosome.
   d.   P and Q are never transferred together. •
   e.   P and Q must be close to the centromere.

12.   A RAPD analysis of a culture arising from a yeast ascospore amplified 4 different-sized bands. These bands
   a.   are from 4 different chromosomes.
   b.   are from 2 homologous chromosome pairs.
   c.   are from four different chromosomal loci. •
   d.   are from two different chromosomal loci.
   e.   are composed entirely of repetitive DNA.

13.   A RAPD analysis of a *Neurospora* strain 1 revealed 2 bands, which we can label *a* and *b*. A RAPD analysis of strain 2 revealed 1 band, which we can call *c*. Progeny of a cross of 1 × 2 gave 1/2 ascospores with *a* and *b*, and 1/2 with *a* and *c*. This means

    a.   *a* and *b* are closely linked.
    b.   *a* and *c* are closely linked.
    c.   *b* and *c* are closely linked. •
    d.   *a* and *b* and *c* are on the same chromosome.
    e.   all RAPD markers are on different chromosomes.

14.   One specific pair of primers revealed a pair of microsatellite repeat bands *p* and *q* in a certain woman; *p* was the larger. The same primers revealed a different pair of bands *r* and *s* in her husband; *r* was larger than *p* and *s* was smaller than *q*. This couple's children can be

    a.   only *p q* or *r s*.
    b.   only *p r* or *q s*.
    c.   only *p s* or *r q*.
    d.   *p r* or *p s* or *q r* or *q s*. •
    e.   lacking all four bands.

15.   A minisatellite marker band present in a mother
    a.   must be present in all children.
    b.   cannot be present in any of her children.
    c.   will be rare in her children.
    d.   will be in 1/4 of her children on average.
    e.   will be in 1/2 of her children on average. •

Open-Ended Questions

1.   Four cloned human genomic DNA fragments (A through D) were tested for sequence-tagged sites (STSs) 1 through 5. The results are shown in the table where + indicates the presence of an STS.

	1	2	3	4	5
A	+	+	+		
B	+		+		
C			+		+
D		+		+	

Draw a contig map of these cloned fragments, showing the relative positions of the STSs.

Solution

```
        4         2         1         3         5
   D_____
          A_____
                B_____
                       C_____
```

2.   A microsatellite probe whose locus shows tight linkage to the locus of the late-onset disorder Huntington's disease (HD) was used to analyze the following family. Autoradiograms using the probe are shown.

Paternal grandfather (HD)	Father (HD)	Mother (unaffected)	Child
—			
		—	—
	—		—
—	—		
		—	

Will the child develop HD later in life?

Solution
Neither of the child's microsatellite alleles is from the grandfather so the child will not have the disease.

3.   Geneticists working on the human genome project were focussing on one band of chromosome 11. They had a collection of eight cloned fragments (clones 1 – 8) that seemed to be from this area. Furthermore they had determined that seven sequence-tagged sites (STSs a – g) were distributed among these clones suggesting clone overlap. The STS content is shown in this table:

Clone	STS content
1	a, b, f
2	a, b
3	f
4	a, c, f
5	c, d
6	e
7	g
8	e, g

Arrange these clones into an overlapping group (contig) showing the regions of overlap and the positions of the STSs.

Solution

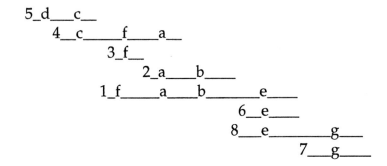

4.   A pair of PCR primers in unique genomic DNA is designed to amplify a microsatellite region. In a married couple this pair of primers amplifies bands of 1.2 and 1.4 kb in the husband, and bands of 0.8 and 1.3 kb in the wife. What offspring genotypes are expected in regard to amplification by these primers, and in what proportions?

Solution
The microsatellite primers amplify one microsatellite repeat locus only, so the bands will act like alleles of one gene and the cross is 1.2/1.4 × 0.8/1.3 and there will be equal numbers of 1.2 /0.8, 1.2/1.3, 1.4 /0.8, and 1.4 /1.3.

5.   A minisatellite probe is used to prepare DNA fingerprints of a man and his wife. The man shows 5 bands of various sizes and the woman 7 bands, some of which are the same as and some are different from her husband's. One of their children is also tested and he shows none of his parents' bands. Is this possible? Explain your answer.

Solution
If the parental markers are all presence/absence heterozygotes then the cross might be for example

1/0 2/0 3/0 4/0 5/0 × 6/0 7/0 1/0 2/0 5/0 8/0 9/0

If so, a child could in theory lack all markers, but this would be rare.

C H A P T E R

18

Control of Gene Expression

Multiple-Choice Questions

1. In the *lac* operon which regions code for no protein?
   a. *z* and *y*
   b. *o* and *z*
   c. *p* and *o* •
   d. *p* and *z*
   e. *o* and *z*

2. The *lac* repressor bind to
   a. lactose and DNA. •
   b. RNA polymerase.
   c. RNA polymerase and DNA.
   d. promoter ans lactose.
   e. beta galactosidase, permease, and transacetylase.

3. In the translation of the *lac* structural genes ribosomes come off the mRNA
   a. after the beta galactosidase and the transacetylase genes only.
   b. only at the end of the operon.
   c. after every protein-coding gene. •
   d. at the operator.
   e. only when there is a mutation.

4.   RNA polymerase binds to
     a.   repressor gene.
     b.   promoter. •
     c.   operator.
     d.   permease gene.
     e.   lactose.

5.   Which of the following is *not* translated from an mRNA molecule
     a.   repressor
     b.   RNA polymerase
     c.   permease
     d.   transacetylase
     e.   lactose •

6.   Beta galactosidase converts
     a.   glucose to galactose.
     b.   galactose to lactose.
     c.   galactose to glucose and lactose.
     d.   lactose to galactose and glucose. •
     e.   lactose to two glucose molecules.

7.   A cell with a polar mutation in beta galactosidase will produce
     a.   acetylase and permease only.
     b.   acetylase only.
     c.   no beta galactosidase, permease or acetylase. •
     d.   beta galactosidase, permease and acetylase.
     e.   no lactose.

8.   A null repressor mutation (I^-) results in
     a.   no transcription.
     b.   inducible transcription.
     c.   transcription but no translation.
     d.   no translation.
     e.   constitutive transcription. •

9.   A promoter mutation (P^-) results in
     a.   no transcription. •
     b.   inducible transcription.
     c.   transcription but no translation.
     d.   no translation.
     e.   constitutive transcription.

10. A super-repressor mutation (I^S) results in
    a. no transcription. •
    b. inducible transcription.
    c. transcription but no translation.
    d. no translation.
    e. constitutive transcription.

11. A partial diploid of genotype $I^- P^+ O^+ Z^+ / I^+ P^+ O^+ Z^-$ will show
    a. inducible production of repressor.
    b. inducible production of beta galactosidase. •
    c. constitutive production of beta galactosidase.
    d. no production of beta galactosidae.
    e. constitutive production of lactose.

12. A partial diploid of genotype $I^S P^+ O^+ Z^+ / I^+ P^+ O^+ Z^-$ will show
    a. inducible production of repressor.
    b. inducible production of beta galactosidase.
    c. constitutive production of beta galactosidase.
    d. no production of beta galactosidae. •
    e. constitutive production of lactose.

13. A partial diploid of genotype $I^+ P^+ O^+ Z^+ Y^- / I^+ P^- O^+ Z^+ Y^+$ will show
    a. inducible production of beta galactosidase. •
    b. inducible production of beta galactosidase and permease.
    c. constitutive production of beta galactosidase.
    d. constitutive production of beta galactosidase and permease.
    e. no beta galactosidase or acetylase production at all.

14. Which of the following does not bind to DNA?
    a. RNA polymerase
    b. DNA polymerase
    c. repressor
    d. beta galactosidase •
    e. CAP-cAMP complex

15. Which of the following is not involved with initiation of transcription in human genes?
    a. TATA-binding protein
    b. RNA polymerase
    c. DNA polymerase •
    d. activators
    e. coactivators

Open-Ended Questions

1.   In each strain shown below determine whether or not there would be expression of β-gal or permease from the *lac* operon. (The galactosidase and permease mutants are not polar.) Show expression with a +, no expression with a – sign.

		Lactose present		Lactose absent	
		β-gal	permease	β-gal	permease
a.	F′ i^- p^+ o^c z^+ y^-/ i^+ p^+ o^+ z^- y^+				
b.	F′ i^+ p^+ o^+ z^+ y^-/ i^- p^+ o^c z^- y^+				
c.	F′ i^s p^+ o^+ z^+ y^+/ i^+ p^+ o^+ z^- y^+				
d.	F′ i^+ p^+ o^c z^+ y^-/ i^s p^+ o^+ z^- y^-				
e.	F′ i^+ p^- o^c z^+ y^+/ i^- p^+ o^c z^- y^+				
f.	F′ i^s p^- o^+ z^+ y^+/ i^- p^+ o^+ z^+ y^+				

Solution

	β-gal	permease	β-gal	permease
a.	+	+	+	–
b.	+	+	–	+
c.	–	–	–	–
d.	+	–	+	–
e.	–	+	–	+
f.	–	–	–	–

2.   In an operon involved in maltose metabolism three mutants were found in the control regions. They were called *a*, *b*, and *c*. Constitutive expression occurred from the a^- and c^- mutants, whereas the b^- mutant resulted in no expression, even in the presence of maltose. When an F′ containing a^+ was introduced into an a^- strain it still maintained

constitutive expression. An F' b^+ in a b^- strain still showed no expression either with or without maltose. Only when an F' c^+ was introduced into the c^- strain did a change in expression occur. The F' c^+/c^- strain now had expression only in the presence of maltose. From this information determine which of the control regions are involved in the a^-, b^-, and c^- mutants.

Solution
Initially we can deduce that both the a and c mutants are either o^c or i^-, and the b mutant must be caused by a promoter mutation. The F' analysis shows that a must be the operator locus because the c^+/c^- partial diploid is now inducible. The c locus must be the repressor.

3. In a bacterial operon similar to the *lac* operon, the R gene codes for a repressor, P is a promoter, O an operator, and A a structural gene for one of the enzymes in the operon. The allele R^- represents lack of repressor, R' represents inability to bind to the inducer, P^- cannot bind RNA polymerase, O' cannot bind the repressor, and A^- is a mutation leading to lack of enzyme activity. In the following genotypes, determine if the cell will show enzyme A activity in the absence and in the presence of the inducer.

    a. $R^+ P^- O^+ A^+$
    b. $R^+ P^+ O' A^+$
    c. $R^- P^+ O^+ A^+$
    d. $R^- P^+ O^+ A^-$
    e. $R^+ P^+ O' A^-$
    f. $R' P^+ O^+ A^+$
    g. $R' P^+ O' A^+$
    h. $R' P^- O' A^+$
    i. $R^+ P^+ O^+ A^+$

Solution

(Result without the inducer is written first.)

a.	no	no
b.	yes	yes
c.	yes	yes
d.	no	no
e.	no	no
f.	yes	yes
g.	no	no
h.	no	yes

4.   In the table below state whether or not the genetic regions of the *lac* operon listed are transcribed in wild-type cells of *E. coli*.

	Region	Transcribed in	
		Presence of lactose	Absence of lactose
a.	*z* gene		
b.	*I* gene		
c.	*lac* promoter		
d.	*y* gene		

Solution

a.	yes	no
b.	yes	yes
c.	no	no
d.	yes	no

5.   In the following six partial diploids, state whether beta galactosidase activity will be inducible, noninducible, or constitutive. Explain your answers.

a. $i^+ o^c z^+ / i^s o^+ z^+$

b. $i^+ o^+ z^+ / i^c o^+ z^+$

c. $i^s o^c z^- / i^c o^+ z^+$

d. $i^c o^+ z^+ / i^+ o^c z^-$

e. $i^+ o^+ z^- / i^s o^c z^+$

f. $i^+ o^+ z^+ / i^s o^+ z^-$

Solution

a.   constitutive

b.   inducible

c.   noninducible

d.   inducible

e.   constitutive

f.   noninducible

C H A P T E R

19

Mechanisms of Genetic Change I: Gene Mutation

Multiple-Choice Questions

1. GC -> CG mutation is called
    a.   nonsense.
    b.   missense.
    c.   frameshift.
    d.   transition.
    e.   transversion. •

2. GC -> AT mutation is called
    a.   nonsense.
    b.   missense.
    c.   frameshift.
    d.   transition. •
    e.   transversion.

3.   Site-directed mutagenesis is used to change a codon from AAA lysine to AGA arginine but the phenotype produced is still wild-type. This type of mutation is called
   a.   suppression.
   b.   neutral. •
   c.   nonsense.
   d.   missense.
   e.   frameshift.

4.   A prototrophic yeast strain is obtained after mutational treatment of an auxotroph. The prototroph is crossed to a wild-type strain (also prototrophic). Some progeny were auxotrophix. The change to prototrophy was caused by
   a.   back mutation.
   b.   suppression. •
   c.   nonsense mutation.
   d.   frameshift mutation.
   e.   missense mutation.

5.   A mutation results in an abnormally short protein. The mutation was most likely of a type called
   a.   missense.
   b.   nonsense. •
   c.   antisense.
   d.   frameshift.
   e.   deletion.

6.   In *E. coli* a region of a gene with repeats of the sequence CTGG will be prone to
   a.   reversion.
   b.   missense mutation.
   c.   nonsense mutation.
   d.   frameshift mutation. •
   e.   amber mutation.

7.   In *E. coli* a region flanked by two repeats of a sequence such as GTGGTGAA is prone to
   a.   deletion. •
   b.   missense mutation.
   c.   duplication.
   d.   inversion.
   e.   frameshift mutation.

8. The rare enol form of thymine pairs with guanine. If a thymine enolization accurs during replication, what would be the mutational event?
    a. CG to AT
    b. GC to TA
    c. AT to TA
    d. TA to CG •
    e. CG to GC

9. Which process is not a major source of spontaneous mutations?
    a. tautomerization
    b. depurination
    c. deamination
    d. dimerization •
    e. enolization

10. Fragile X syndrome is caused by
    a. trinucleotide repeats. •
    b. free radicals.
    c. microdeletions.
    d. 5-bromouracil.
    e. depurination.

11. After mutagen treatment, a molecule of 2-aminopurine (an adenine analogue) incorporates into DNA. During replication the 2-AP protonates. The mutational event caused by this will be
    a. AT to CG.
    b. GC to AT.
    c. AT to TA.
    d. AT to GC. •
    e. GC to CG.

12. During mutagenic treatment with nitrous acid, an adenine deaminates to form hypoxanthine which bonds like guanine. The mutational event would be
    a. AT to GC. •
    b. AT to CG.
    c. AT to TA.
    d. GC to AT.
    e. GC to TA.

13. One of the following is not a mutagen
    a. 5-bromouracil
    b. proflavine
    c. acridine orange
    d. ethyl methane sulfonate
    e. deoxy-adenosine monophosphate •

14.   A missense mutation in *Neurospora* will revert by treatment with nitrous acid, but not by hydroxylamine. The original mutation (not the reversion) must have been
    a.   AT to GC.
    b.   AT to TA.
    c.   AT to CG.
    d.   GC to AT. •
    e.   GC to TA.

15.   In *E. coli* mutation arising during repair is mostly by
    a.   thymine dimer splitting.
    b.   excision repair.
    c.   mismatch repair.
    d.   recombinational repair.
    e.   SOS repair. •

Open-Ended Questions

1.   PCR was used to amplify a small region at the 5′ end of the coding region of three mutant alleles in yeast. This region was sequenced as follows, grouped in codons.

Wild-type	5′ GAACTCGAGCTAAAT	3′
Mutant 1	5′ GAACTCGAGCTAATT	3′
Mutant 2	5′ GAACTCAAGCTAAAT	3′
Mutant 3	5′ GAACTCGAGCCTAAAT	3′

Classify these mutations into as many categories as possible.

Solution
Mutant 1 shows A to T, so it is an AT to TA transversion, and missense.
Mutant 2 shows G to A so it is a GC to AT transition, and also missense.
Mutant 3 is a frameshift mutation caused by insertion of a C.

2.   Three mutations were obtained in a bacterial gene for whose protein product an antibody is available. Both Northern and Western (protein/antibody) analyses were performed on the mutants, and the results are summarized below.

Northern				Western			
+	1	2	3	+	1	2	3
−	−	−					
					−		
					−	−	

What types of mutations are these?

Solution
Mutant1, Missense or frameshift (no change in RNA or protein size)
Mutant 2, Nonsense (RNA same size, but protein shorter)
Mutant 3, Possibly a promoter mutation as no RNA is transcribed; or possibly a deletion of the gene.

3.   Two new hair dyes *A* and *B* were tested in the Ames test. Both were tested with and without liver homogenate. Two auxotrophic strains of bacteria were used, one in which the auxotrophy was caused by nonsense mutation (mutant 1), and in the other by a frameshift (mutant 2). After treatment the cells were plated on minimal medium. The + signs show experiments in which prototrophic colonies were observed. Explain the mutagenic action of these chemicals, and say which would probably be safe for human use.

	1		2	
Dye	No liver	Liver	No liver	Liver
A	–	+	–	–
B	–	–	+	–

Solution
A causes base-pair substitutions but only after liver activation. *B* causes frameshifts but the mutagenic action is inactivated by the liver extract. *B* would probably be safer than *A*, but more tests would be needed before use in human populations.

4.   A *Neurospora* nonsense mutation known to be UAG is treated with two compounds, hydroxylamine and 5-bromouracil. Do you expect either of these agents to induce revertants? Explain.

Solution
HA causes only GC to AT changes so can only produce changes of the type UAG to UAA, which is also a nonsense mutation. However, %BU can cause AT to GC and GC to AT changes, therefore either of the first two codon positions can be changed to GC (or CG) producing missense mutations. hence 5BU reverts, but Ha doesn't.

5.   A human gene K contains a single restriction site for *Eco*RI. A common recessive mutant allele that results in kidney disease has a base-pair substitution that inactivates this site. How could this knowledge be used in diagnosing the mutant allele in heterozygotes?

Solution
There must be *Eco*RI sites outside the gene so a the wild-type allele would be cut into two fragments whereas the mutant only one. If a probe for gene is available that spans the RFLP site, Southerns of heterozygotes would show three bands, and homozygous wild-types just one.

6. Sequential mutations affecting one codon caused the changes

arg -> thr -> met -> leu -> gln

What base pair changes produced these amino acid changes?

Solution
  AGG -> ACG -> AUG -> CUG -> CAG

7. Five different mutations were derived from base pair substitutions at one single codon. In this codon the mutant alleles had arginine, leucine, glycine, serine, and cysteine. What was the wild-type codon?

Solution
Tryptophane (UGG) can give all five amino acids by single base pair transitions or transversions at positions 1, 2, or 3.

8. a. A *met*+ strain of *Neurospora* was treated with a mutagen to create *met*- mutants. The mutants were reverted to *met*+ with HA (hydroxylamine). What was the original mutation on the molecular level? Name a possible mutagen that could have been used to cause the original mutation.

b. Two of the revertants showed odd results when crossed with a WT *met*+ strain.

met+ revertant A × *met*+ WT ——> 88% *met*+
                                                  12% *met*-

met+ revertant B × *met*+ WT ——> 93% *met*+
                                                  7% *met*-

Explain using diagrams how these results occurred and why the above numbers of offspring were obtained.

Solution
  a. HA causes GC to AT transitions so the original mutation must have been AT to GC, possibly caused by nitrous acid.
  b. Reversion is not true reversion but suppression. Furthermore, in case A the suppressor must be 24 m.u. away from the *met* locus, although in B it is 14 m.u. away.

20

Mechanisms of Genetic Change II: Recombination

Multiple-Choice Questions

1. In a haploid cross of $m \times +$, which of the following linear octads shows the occurrence of gene conversion?
    a.   ++++*mmmm*
    b.   ++*mmmm*++
    c.   +++++*mmm* •
    d.   *mm*++++*mm*
    e.   *mmmm*++++

2. In a cross $m \times +$, which of the following asci reveals the presence of a heteroduplex DNA spanning the mutant site, <u>without</u> associated gene conversion?
    a.   +++++*mmm*
    b.   *mmmmm*++
    c.   *mmmmmm*++
    d.   ++++++*mm*
    e.   +++*m*+*mmm* •

3. Which of these processes is not part of a molecular model for crossing over?
   a. strand breakage
   b. strand invasion
   c. ligation
   d. branch migration
   e. thymine dimerization •

4. In a fungal cross between two alleles *m1* and *m2* of a certain gene, which of the following asci shows co-conversion at both mutant sites?
   a. $m1 +/+ m2/+ m2/+ m2$ •
   b. $m1 +/+ +/+ m2/+ m2$
   c. $m1 +/m1 +/+ m2/+m2$
   d. $m1 +/m1 +/+ m2/+ m2$
   e. $m1 +/m1 m2/ + +/+ m2$

5. According to the Holliday model, which of the following linear ascus ratios is caused by correction of both hetroduplexes?
   a. 5:3
   b. 6:2 •
   c. 3:1:1:3
   d. 4:4
   e. 2:2:2:2

6. In the Holliday model, which of the following comes last in recombination?
   a. heteroduplex formation
   b. strand breakage
   c. excision repair •
   d. DNA unwinding
   e. recognition of fixed break point

7. In the Meselson–Radding model, which of the following comes last?
   a. strand invasion
   b. branch migration
   c. chain removal
   d. resolution •
   e. ligation

8. In a cross of $+ \times +$, which of the following is false?
   a. gene conversion cannot be detected
   b. no aberrant ratios can be detected
   c. no half-chromatid crossovers are produced •
   d. no crossovers can be detected
   e. biparental DNA duplexes form

9.   In a homoallelic cross *arg* × *arg* is a linear ascus of the type
      arg / arg / arg / + / arg / arg / arg / arg
This is an example of
    a.   half chromatid conversion.
    b.   whole chromatid conversion.
    c.   strand migration.
    d.   coconversion.
    e.   reversion. •

10.   In the cross

$$\frac{A \quad 1 + \qquad B}{a \quad + \ 2 \qquad b} \quad \text{by}$$

1 and 2 are nonrevertible mutant sites of the *meth2* gene, and the *A* and *B* loci are closely-linked flanking markers. Haploid progeny ascospores of the genotype *a met$^+$ b* are examples of
    a.   mutation.
    b.   interalleleic complementation.
    c.   gene conversion at site 1.
    d.   gene conversion at site 2. •
    e.   nonreciprocal recombination.

11.   Gene conversion occurs only
    a.   around the centromere.
    b.   at a crossover site. •
    c.   between gene loci.
    d.   between sister chromatids.
    e.   near heterochromatin.

12.   Which of the following enzymes is not used in molecular recombination?
    a   telomerase •
    b   ligase
    c   DNA polymerase
    d   endonuclease
    e   exonuclease

13.   A Holliday junction is composed of
    a.   chromatids.
    b.   DNA double helices.
    c.   chromosomes.
    d.   polynucleotide strands. •
    e.   centromeres.

14. The substrate for the gene conversion process is thought to be
   a.   Holliday junctions.
   b.   DNA breaks.
   c.   two paired strands of 5'-3' polarity.
   d.   mispaired nucleotides. •
   e.   unequal crossovers.

15. Which of the following is not a feature common to the Holliday, Meselson–Radding, and double-strand break models of molecular recombination?
   a.   strand invasion •
   b.   phosphodiester bond breakage
   c.   DNA polymerase
   d.   heteroduplex formation
   e.   ligase action

Open-Ended Questions

1.   In a cross of a wild-type *Neurospora* strain to a strain requiring pantothenic acid as a result of a mutation in the *pan-2* gene, 130 asci show ratios of 4 + : 4 *pan-2*, but one ascus showed a ratio of 5 + : 3 *pan-2*. How would the Holliday model explain this rare exceptional ascus?

Solution
A pair of heteroduplexes formed spanning this locus. Gene conversion occurred in one of these heteroduplexes in the direction pan->+, and there was no gene conversion in the other hetroduplex. Hence the four DNA molecules were

```
++    +m   +m   mm
 1     2
heteroduplexes

correction at 1
```

++   ++   +m   mm

replication

2.   In a fungal cross *his-2 leu2* + × + + *nic3*, all three loci are linked quite closely in the order shown. An asci analysis was carried out, and all ascospores were tested for leucine requirement One rare ascus was obtained that showed a ratio of 5 *leu2* : 3 +. This was thought to be an example of gene conversion, but a sceptic might say that there was a mistake in ascospore isolation and one of the ascospores was from another ascus. What test would you do to rule out this trivial explanation?

Solution
Test all the ascospores of the rare ascus for histidine and nicotinamide requirement; they should show 4:4 ratios because gene conversion would be unlikely to occur at all loci at once.

3. How do the Holloday and Meselson–Radding models differ in their explanations of the 6:2 aberrant ascus ratio?

Solution
Holliday's model invokes two heteroduplexes which both correct the same way, mutant to wild-type or wild-type to mutant. The Meselson–Radding model just needs one heteroduplex, with correction to the majority allele (the type that ends up as the '6').

4. If at a mutant site within a gene there is a GC nucleotide pair, and the equivalent site in wild-type is AT, mismatched heteroduplexes of the type GT and AC will form. If the correction mechanism preferentially excises the wild-type nucleotide in both cases, which aberrant octad ratios are expected to be favored?

Solution
Aberrant ratios in which the mutant is in the excess will be favored. So,
     6m:2+ >> 6+:2m
     5m:3+ >> 5+:3+

5. In the ascomycete fungus *Ascobolus* (which forms octads), a cross is made + × *t*, where *t* gives tan ascospores in contrast to the black of wild-type. In addition both parents carry null mutations for a gene coding for a function specific to heteroduplex correction. If heteroduplexes occur regularly at the tan locus, what aberrant ratios are expected to be most common?

Solution
If two heteroduplexes form, then the resulting asci will be an aberrant 4:4 or (3:1:1:3). If only one heteroduplex forms at the tan locus, then the ratio will be 5:3.

6. In *Neurospora* the cross is made

$$\frac{ylo \quad pan\text{-}2^2 \quad tryp}{+ \quad\quad pan\text{-}2^1 \quad\quad +} \quad\times$$

(*pan-2^1* and *-2^2* are alleles of the *pan-2* locus, and *ylo* and *tryp* are closely-linked flanking markers; *ylo* = yellow color, and *tryp* = trypotophane requirement.)
Ascospores were plated on minimal medium plus tryptophane and rare *pan-2$^+$* colonies grew. These were isolated and tested further and the two most common flanking marker combinations were found to be *ylo tryp$^+$* and *ylo tryp*, and the *ylo$^+$ tryp$^+$* and *ylo$^+$ tryp* combinations were relatively rare. Explain these results in terms of polarity in molecular recombination.

Solution

The sites must have been 1 to the left and 2 to the right, and gene conversion occurred more commonly at site 2 (at the right). Therefore heteroduplex DNA forms more commonly at the right hand site 2 because of polarity. The *ylo tryp*⁺ combination arises from conversion to + at site 2, and the Holliday junction resolved into a full crossover. The *ylo tryp* combination is also from conversion to + at site 2 but with the Holliday junction resolved into a noncrossover conformation. (Both the rarer combinations require gene conversion to + at the site 1.)

21

Mechanisms of Genetic Change III: Transposable Genetic Elements

Multiple-Choice Questions

1. In the galactose operon, the order of genes is promoter-epimerase-transferase-galactokinase. An IS element in the galactokinase gene will knock out function
   a. only in galactokinase. •
   b. only in epimerase.
   c. only in transferase.
   d. in galactokinase and transferase and epimerase.
   e. in galactokinase and transferase.

2. Hybridization of single-stranded wild-type DNA with DNA from mutations caused by IS elements characteristically show (through electron microscopy)
   a. chi structures.
   b. unpaired tails.
   c. a loop representing IS DNA. •
   d. a loop reprsenting wild-type DNA.
   e. theta structures.

3.   IS-induced mutations are different from missense mutations in that they are
     a.   nonrevertable.
     b.   less severe.
     c.   more severe.
     d.   non-polar.
     e.   not reverted by mutagens. •

4.   Which is a characteristic of a bacterial insertion sequence?
     a.   size approximately 10 nucleotide pairs
     b.   only one copy per genome
     c.   found on bacterial chromosome and not plasmids
     d.   has inverted repeats •
     e.   cannot move to a new locus

5.   Which of the following functions is not carried on plasmids?
     a.   drug resistance
     b.   fertility
     c.   camphor metabolism
     d.   replication origin
     e.   ribosomal RNA •

6.   If DNA from a bacterial culture carrying the *amp*^{R} gene in Tn3 is used to transform a recipient culture of genotype *su*^{R} *sm*^{R}, then the cells are plated on ampicillin, which of the following genotypes could not be recovered from the plates?
     a.   *amp*^{S} *su*^{R} *sm*^{R} •
     b.   *amp*^{R} *su*^{R} *sm*^{R}
     c.   *amp*^{R} *su*^{R} *sm*^{S}
     d.   *amp*^{R} *su*^{S} *sm*^{S}
     e.   *amp*^{R} *su*^{S} *sm*^{R}

7.   Bacterial transpososn structure can be thought of as
     a.   IS sequences flanked by inverted drug resistance genes.
     b.   drug resistance gene(s) flanked by IS elements. •
     c.   drug resistance gene(s) flanked by a pair of mu phage.
     d.   mu phage flanked by two IS elements.
     e.   resistance gene(s) flanked by inverted resistance transfer factors (RTFs).

8. A bacterial cell contains a single copy of a transposon in which one strand carries a G at one site (the wild-type base at this site) and the other strand carries a T (which confers a mutant phenotype). If this cell is allowed to divide through several divisions, but the transposon transposes at the first division through replicative transposition, the colonies derived from these cells will be
    a. mosaic, with wild-type and mutant sectors.
    b. all mutant.
    c. all wild-type.
    d. all GT heteroduplexes.
    e. some wild and some mutant. •

9. Which of the following is <u>not</u> a feature that bacterial and corn transposons have in common?
    a. they both cause unstable mutations
    b. they both carry drug resistance genes in natural populations •
    c. they both have inverted repeats
    d. they both move to new loci
    e. they both cause rearrangements

10. In *Drosophila*, hybrid dysgenesis is caused by
    a. Ty elements.
    b. *copia* elements.
    c. *copia*-like elements.
    d. P elements. •
    e. FB elements.

11. Which of the following is not a feature of hybrid dysgenesis in *Drosophila*?
    a. sterility
    b. high mutation rate
    c. high frequency of chromosome aberrations
    d. high frequency of nondisjunction
    e. high frequency of crossing-over •

12. Transposons are useful in genetic analysis because of their use in
    a. gene tagging. •
    b. preventing crossing over.
    c. stabilizing aneuploids.
    d. generating duplications.
    e. preventing reversion of mutant alleles.

13. Retroviruses replicate using
    a. DNA polymerase.
    b. RNA polymerase.
    c. restriction endonuclease.
    d. reverse transcriptase. •
    e. topoisomerase.

14. Retrotransposons move via an intermediate that is
    a. a double-stranded DNA lollipop.
    b. a retrovirus.
    c. double-stranded RNA.
    d. single-stranded DNA.
    e. single-stranded RNA. •

15. A corn plant is homozygous for a mutant allele that results in no color in the seed (white). The mutant is caused by *Ds* insertion that often exits late in seed development, and there is an active *Ac* element in the genome. The seeds will be
    a. white.
    b. pigmented all over.
    c. white with small spots of pigment. •
    d. white with large spots of pigment.
    e. weakly pigmented.

Open-Ended Questions

1. At the *E. coli lac* operon, a mutant arose that mapped in the beta-galactosidase gene near the operator, but this mutant lacked not only beta-galactosidase activity but also acetylase and permase.
    a. Propose two types of mutational events that could account for this type of mutation.
    b. The reversion rate to wild-type was the same in the presence or absence of proflavine. Does this allow you to distinguish your two alternatives?

Solution
    a. 1. Transposon or IS insertion giving a polar mutation affecting genes translationally downstream; 2: a +/- frameshift mutation.
    b. Since proflavine causes +/- mutations it would be expected to revert a frameshift mutation and so the mutation is probably caused by an insert of some kind – IS or Tn.

2. A corn plant carries mutant alleles for the single gene that deposits colored pigment in leaves. (When active this gene deposits purple pigment which, with the green of chlorophyll, makes the plant brown.) The plant in question is green with brown spots all over it. When the plant was selfed 1/4 of the progeny were plain green, and 3/4 were green with brown spots.
    a. Explain the spotted phenotype of the original plant.
    b. Explain the 3:1 ratio.
    c. Could you ever obtain a fully brown progeny from this self?

Solution

a.   An active transposon had entered the gene that causes pigment deposition, thereby inactivating it. Frequent excisions caused pigmented spots, one from each excision.

b.   The plant was heterozygous in some way. Perhaps it was $a^u a$ where a^u is an unstable allele caused by Tn insertion, and a is a stable allele caused by a base pair change. Then 3/4 of the progeny will be a^u- and spotted.

c.   Yes, if a reverted, or if excision occured in the germinal tissue.

3.   A yeast strain of wild-type phenotype carried a single copy of an active transposon. This strain was plated and one rare colony on the plate was extremely small; subsequent studies showed that it had very thin cell walls. A Southern analysis revealed a new fragment hybridizing the transposon probe, showing that the small strain now had two copies of the transposon. The new fragment was extracted and using the know transposon sequence, a primer was designed to sequence outwards from the transposon tip. The sequence obtained was part transposon DNA but when the next 500 nucleotides were tested against the DNA data bank, its putative amino acid sequence was found to be glucose-6-phosphate dehydrogenase. Explain how the mutant phenotype arose, and the significance of the G6PD.

Solution

From its original benign location (presumably not inside a gene) a copy of the transposon had moved into the G6PD gene, inactivating it. The sequence was a junction between the Tn and the gene. This enzyme, normally active in carbohydrate metabolism, when inactive through mutation can apparently interfere with cell wall synthesis (polysaccharide synthesis).

4.   A gene *G*, containing two small introns, was inserted into a yeast transposon at a position where it did not interfere with transposition. The transposon moved to new locations, and the structure of the transposon in its new locations was studied. It was found that at each new location the transposon still contained gene *G* but without introns in every case. Explain the loss of the introns, and state what this tells us about the mechanism of transposition.

Solution

The transposon must be a retrotransposon, which moves via an RNA intermediate. The introns must have been spliced out of the RNA, and then reverse transcriptase made DNA copies which were inserted into the new locations.

5.   In *Neurospora*, the cloned *am* (amination) gene hybridizes to a single 4 kb *Hae*III genomic fragment in a wild-type strain from Africa. Three *am*⁻ mutants arose spontaneously in this strain. The probe was used on *Hae*III digests of these three strains and in two of them the probe bound to a single 9 kb fragment and in the third one the probe bound to two fragments of 3 kb and 1 kb in size. What is likely to be the nature of these three mutations?

Solution
The types with 9 kb fragments are probably caused by the insertion of a 5 kb transposon into the *am* gene. The other mutation is probably a spontaneous missense mutation that happened to create an *Hae*III site within the *am* gene.

6.   In corn, two independent genes are involved in kernel color. When both dominant alleles are present (*CCRR*) the color is brown, when either is recessive the color is yellow, and if both are recessive (*ccrr*) the kernels are white. A brown corn plant from a pure line, known to have unstable color, is crossed with a *ccrr* tester (with stable color). Below are the gametes from the brown plant, describe the potential kernel colors in each case.
   a.   *Ds* on a different chromosome from both *C* and *R*, no *Ac*.
   b.   *Ds* and *Ac* on *C* chromosome.
   c.   *Ds* on *R* chromosome, *Ac* on another.
   d.   *Ds* in *R* gene, *Ac* in *C*.
   e.   *Ac* on *C* chromosome, no *Ds*.
   f.   *Ac* in *C* gene, no *Ds*.

Solution
   a.   *Ds* cannot move, so plant is brown.
   b.   *Ds*-induced breaks might cause loss of part of chromosome containing *C* leaving only *R* active; yellow spots on brown background.
   c.   *Ac* transposase will cause *Ds* breaks which might cause loss of *R*-containing part of chromosome; also yellow spots on brown.
   d.   Both *Ds* and *Ac* can exit; some spots yellow (exit from *C* or *R*), and some brown (exit from both) all on brown background.
   e.   *Ac* can cause breaks; yellow spots on brown.
   f.   Yellow background, exit from *C* cause brown spots.

C H A P T E R

22

The Extranuclear Genome

Multiple-Choice Questions

1. In animals and fungi, extranuclear genes are in the
    a. endoplasmic reticulum.
    b. golgi bodies.
    c. chloroplasts.
    d. lysosomes.
    e. mitochondria. •

2. Which of the folowing is not in the cytoplasm
    a. golgi body
    b. nucleolus •
    c. mitochondrion
    d. lysosome
    e. secretory vesicle

3. Green/white mosaics in plants are often caused by mutations in
    a. mitochondria.
    b. chloroplasts. •
    c. viruses.
    d. plasmids.
    e. prions.

4.   In a cross made with variegated four o'clock plants, pollen is provided from flowers on a green sector, and ovules from a white sector. The progeny are expected to be
    a.   all green.
    b.   all white. •
    c.   all variegated.
    d.   some variegated, some green, some white.
    e.   some variegated, some white.

5.   In four o'clock plants that show green/white sectoring, the mechanism for producing sectors is best described as
    a.   mitotic crossing over.
    b.   meiotic crossing over.
    c.   unstable mutations in cpDNA.
    d.   cytoplasmic sgeregation. •
    e.   complementation.

6.   In the haploid fungus *Neurospora*, a strain with the mutant phenotypes poky (mitochondrial) and *his-2⁻* (nuclear) is crossed as maternal parent to a wild-type strain of opposite mating type acting as paternal parent. What percentage of progeny will be poky *his-2⁺*?
    a.   0
    b.   25
    c.   50 •
    d.   75
    e.   100

7.   In the haploid fungus *Neurospora*, a strain with the mutant phenotype poky (mitochondrial) is crossed as paternal parent to a wild-type strain of opposite mating type acting as maternal parent. What proportion of progeny will be poky?
    a.   0 •
    b.   0.25
    c.   0.50
    d.   0.75
    e.   1.0

8. In the haploid fungus *Aspergillus*, a new mutation arose causing slow growth, named '*sg*'. It was supected that *sg* was a mitochondrial mutation so a heterokaryon test was performed. A heterokaryon was made between a strain that was *sg* and also carried the nuclear mutation *arg1*, and another strain that carried the nuclear mutation *leu2*. Then conidial isolates from the heterokaryon were tested. If *sg* was in fact a mitochondrial mutation, this would be demonstrated by finding that there were many conidial isolates of the genotype

    a.   *sg arg1.*
    b.   *sg leu2.* •
    c.   *sg arg$^+$ leu$^+$·*
    d.   *sg arg1 leu2.*
    e.   fully wild-type.

9. In the snail *Limnaea*, shell coiling is determined by the nuclear alleles S (dextral) and s (sinistral). There is a maternal effect in which progeny phenotype is determined by the genotype of the mother. A sinistral snail was selfed and all the progeny were dextral. When these dextral types were selfed it was found that 3/4 of the dextrals had dextral progeny, but 1/4 had sinistral progeny. The original snail must have been of genotype

    a. SS
    b. ss
    c. Ss •
    d. S
    e. s

10. In *Chlamydomonas* erythromycin, resistance is inherited on cpDNA. A cross is made

    $mt^- ery^R \times mt^+ ery^S$.

What percentage of progeny will be $ery^R mt^+$?

    a.   0 •
    b.   25
    c.   50
    d.   75
    e.   100

11. In yeast the *ery^R* mutation is inherited mitochondrially. A cross is made

eryR × eryS

Diploids are made and diploid buds are allowed to go through meiosis. A tetrad was isolated and the first ascospore tested was *eryR*. What proportion of the remaining three ascospores in this tetrad is expected to be *ery^R* also?

    a.   none
    b.   1/3
    c.   1/2
    d.   2/3
    e.   all of them •

12. In *Saccharomyces cerevisiae*, the mutations *oli^R* and *spi^R* are inherited mitochondrially. A cross is made

oli^R × spi^R

Diploids are made and diploid buds are allowed to go through meiosis. A terad was isolated and the first acospore to be tested was *oli^R spi^R*. What proportion of the remaining ascospores in this tetrad is expected to be *oli^R spi^R*?

    a.   none
    b.   1/3
    c.   1/2
    d.   2/3
    e.   all of them •

13. In *Saccharomyces cerevisiae*, a new mutation resistant to nystatin arose. A grande strain that was *nys^R* was treated with ethidium bromide and many petite mutations were obtained. If the *nys^R* locus is mitochondrial, then the petites are expected to be

    a.   all *nys^R*
    b.   all *nys^S*
    c.   a few will be *nys^R*, the remainder *nys^S*
    d.   a few will be *nys^S*, the remainder *nys^R* •
    e.   approximately equal numbers of *nys^R* and *nys^S*

14. The mtDNA of neutral petite mutants
    a.   has an altered restriction map.
    b.   has an altered base ration.
    c.   has deletions.
    d.   has frameshifts caused by ethidium bromide.
    e.   is absent. •

15. Which of the following types of genes are not in the mtDNA of yeast?
    a. ribosomal RNA
    b. tRNAs
    c. some cytochrome oxidase subunits
    d. some ATPase subunits
    e. some Krebs cycle enzymes •

16. Which of these general categories of genes is not found in the cpDNA of the liverwort *Marchantia*?
    a. transcription
    b. translation
    c. photosynthesis
    d. electron transport
    e. replication •

Open-Ended Questions

1. A plant shows sectors of green and of white in its photosynthetic tissue caused by a mutation in cpDNA. Flowers appear all over the plant, and these are allowed to self-pollinate. Seeds are collected from
    a. flowers in pure white sectors
    b. flowers in pure green sectors
    c. flowers in regions that are mixtures of small green and white sectors.
What type of progeny are expected from these three types of selfs?
    d. If white progeny are predicted, comment on their possible survival.
    e. If white tissue is nonphotosynthetic (which seems likely) how can it suvive to be a white sector?

Solution
Because these mutations are inherited maternally the progeny phenotypes will reflect the tissue the flowers are on.
    a. all progeny white
    b. all progeny green
    c. some white, some green, some variegated.
    d. they will die after exhausting all maternal supplies
    e. the white parts parasitize the green parts.

2. A *Neurospora* heterokaryon is made between the following haploid strains
    Strain 1   *A ad-3 nic-2* poky
    Strain 2   *A pan-2 met-6*
(*A* is an allele of the mating type gene – heterokaryons must be made beween strains of the same mating type. Poky is amitochondrial mutation; all others are nuclear auxotrophic mutations.)

If conidia from this heterokaryon are isolated individually, what genotypes are expected?

Solution
Strain 1: More heterokaryons, probably prototrophic and probably non-poky initially. Because of the suppressiveness of mitochondrial mutaions, the HK might soon become poky.
Strain 2: Segregants of four different types

A *ad-3 nic-2* poky
A *ad-3 nic-2* nonpoky
A *pan-2 met-6* poky
A *pan-2 met-6* nonpoky

(Note: no recombinants of nuclear genes; nuclei do not fuse.)

3.   A strain of *Neurospora* isolated from a certain Hawaiian population showed sickly growth, often ending in growth cessation. This sickly strain showed two outcomes when crossed as maternal parent to two other nonsenescent strains (which we can call 1 and 2) from the same Hawaiian population. The results were as follows.

sickly × nonsickly 1 -> progeny all sickly

sickly × nonsickly 2 -> 1/2 progeny sickly, 1/2 nonsickly

Propose a genetic explanation for the results of these two crosses, showing genotypes of strains under your explanation.

Solution
The sickly phenotype is inherited maternally in cross 1, suggesting it is mitochondrially based, possibly caused by a mitochondrial plasmid. Cross 2 shows segregation of a Mendelian allele that is formally acting as a suppressor of the sickly phenotype, possibly by regulating the action of the plasmid.
Genotypes ([] = mtDNA-based)

[sick] × + -> all [sick]

[sick} × *su* -> 1/2 [sick] +, 1/2 [sick] *su*

4.   The yeast mutations *mit1*, oli^R, and spi^R are all mitochondrially inherited. What tetrads are expected from the crosses
   a.   *mit1* × *mit1*$^+$
   b. oli^R × oli^S
   c. spi^R × spi^S
   d. spi^R oli^S × spi^S oli^R

Solution

    a.   All members of a tetrad will be either *mit1* or *mit1+*.

    b.   All members of a tetrad will be either *oli*R or *oli*S.

    c.   All members of a tetrad will be either *spi*R or *spi*S.

    d.   All members of a tetrad will be either *spi*R *oli*S, or *spi*R *oli*R, or *spi*S *oli*R or *spi*S *oli*S.

5.   The mtDNA of two petites are probed with the large ribosomal RNA gene cloned in a pUC vector. The results are shown in the following diagram. Explain these results and draw a rough map of the mtDNA regions concerned in your explanation.

EcoRI				BglII			
wild-type		petite		wild-type		petite	
gel	Xray	gel	Xray	gel	Xray	gel	Xray
—		—					
				—		—	
				—	—	—	
				—		—	
—		—					
				—		—	
—	—						
—		—					
				—		—	
—	—	—	—				
—		—					
				—		—	
				—	—	—	
				—			
—		—					

Solution

i)   There is an *Eco* site in the middle of the rRNA gene, but no *Bgl* site.

ii)   The petite deletion removed one of the *Eco* fragments, and shortened the *Bgl* fragment.

iii)   Approximate map:

```
B/E_____E_____B_____E
                              //////////
                              rRNA gene
I_____I
              deletion
```

C H A P T E R
23
Gene Regulation during Development

Multiple-Choice Questions

1. Enhancer elements regulate gene expression at the level of
    a. gene structure.
    b. transcription. •
    c. transcript processing.
    d. translation.
    e. activation of protein product.

2. As described in Chapter 23, a chromosomal inversion in *Drosophila* can cause a gain-of-function dominant mutation through
    a. inactivation of the transcription unit of one gene.
    b. deletion of key regulatory elements from one gene.
    c. fusion of the regulatory elements of one gene to the regulatory elements of another gene.
    d. fusion of the regulatory elements of one gene to the transcription unit of another gene. •
    e. none of the above.

3.   The pathway of somatic sex determination and differentiation in *Drosophila* provides examples of regulation at the level(s) of
    a.   gene structure.
    b.   transcription. •
    c.   transcript processing. •
    d.   translation.
    e.   activation of protein product.

4.   Which of the following karyotypes provide evidence that somatic sex determination mechanisms are different in *Drosophila* and mammals?
    a.   XX
    b.   XXY •
    c.   XY
    d.   X0 •
    e.   individuals with a mixture of XX and XY cells •

5.   In *Drosophila*, which of the following genotypes will exhibit male somatic sexual phenotypes?
    a.   an XX individual containing wild-type alleles of all of the genes involved in sex determination
    b.   an XXY individual homozygous for a null mutation of *Sxl* •
    c.   an XX individual containing a nonsense mutation in one of the male-specific exons of the *dsx* gene
    d.   an XY individual containing a nonsense mutation in one of the male-specific exons of the *dsx* gene
    e.   an XY male containing a constitutively active *Sxl* gene

6.   Supporting evidence for the identification of the transcription unit encoding the Y chromosome-linked mammalian sex determining gene *SRY/Sry* includes which of the following?
    a.   The *SRY/Sry* transcription unit is included in all sex reversal duplications of the tip of the Y chromosome.
    b.   XX–XY mosaic individuals are generally phenotypically male.
    c.   The *SRY/Sry* transcription unit includes 7 exons.
    d.   The *SRY/Sry* transcription unit is expressed in the gonadal ridge on the kidney of developing mouse embryos. •
    e.   The *SRY/Sry* transcription unit is included in the 14 kb of mouse genomic DNA that can be transformed into XX mouse embryos and cause them to develop as somatically male mice. •

7. The constant region of each immunoglobulin includes the polypeptide sequences that
    a.   provide diversity to the light chains and heavy chains.
    b.   control antigen–antibody interactions.
    c.   produce disulfide bridges between the subunits of the immunoglobulin tetramer. •
    d.   act as enzymes in the V–D, D–J and V–J site-specific recombination reactions.
    e.   are expressed in T lymphocytes.

8. Antibody diversity arises from which of the following factors?
    a.   multiple V and J segments in light chain genes •
    b.   multiple V, D and J segments in heavy chain genes •
    c.   allelic exclusion
    d.   maturation of a B lymphocyte into a plasma cell
    e.   complete independence of light chain and heavy chain rearrangements in a given cell •

9. The light chain enhancer element induces high levels of transcription in B lymphocytes and plasma cells. The location of this enhancer element near the C segment of the light chain gene
    a.   ensures that any properly rearranged light chain gene can be transcribed. •
    b.   ensures proper splicing of the light chain primary transcript.
    c.   determines the location of DNA rearrangements between the J segments and the C segment.
    d.   all of the above.
    e.   none of the above.

10. Sex determination in *Drosophila* exemplifies a developmental pathway in which
    a.   one developmental state is the default, whereas its alternative requires the activation of a series of regulatory gene products.
    b.   an early decision is 'remembered' through the establishment of an autoregulatory loop.
    c.   regulation occurs at more than one level from the gene to the active protein.
    d.   an "on–off" decision for a master switch is made early in development.
    e.   all of the above •

11. Sexual nonautonomy in mammals reflects the fact(s) that
    a.   many tissues display sexually dimorphic phenotypes.
    b.   all cells of either sex have androgen receptors. •
    c.   besides its role in making sperm, the testis synthesizes male-specific hormones. •
    d.   one set of cells sends instructive signals (hormones) that are received and interpreted by other tissues in the body.
    e.   all of the above •

12. Which of the following are properties of X chromosome dosage compensation in mammals?
    a. Only a single X chromosome remains active in each cell. •
    b. All X chromosomes form Barr bodies.
    c. X inactivation occurs randomly late in development.
    d. X inactivation never occurs in individuals with a Y chromosome, regardless of the number of X chromsomes they possess.
    e. In certain tissues and in certain mammals, parental imprinting causes the paternal chromosome to be preferentially inactivated. •

13. Which of the following *Drosophila* karyotypes would be phenotypically male?
    a. XY with two sets of autosomes •
    b. X0 with two sets of autosomes •
    c. XX with three sets of autosomes
    d. XXY with four sets of autosomes •
    e. XXX with three sets of autosomes

14. In *Drosophila* sex determination, numerator and denominator elements function act as
    a. splicing factors determining the splicing pattern of the *Sxl* gene throughout development.
    b. splicing factors for the *dsx* gene.
    c. transcription factors regulating *Sxl* transcription early in embryogenesis. •
    d. transcription factors for expression of the *dsx* gene.
    e. transcription factors regulating *Sxl* transcription throughout development.

15. Some mammals have one light chain gene and one heavy-chain gene for the production of immunoglobulins. If in such a mammal, the light-chain gene has 300 V segments and 4 J segments, while the heavy-chain gene has 500 V segments, 18 D segments and 5 J segments, the total number of different antibody molecules that could be generated in all possible DNA rearrangements of the 2 genes is
    a. 45,000.
    b. 1,200.
    c. 54,000,000. •
    d. 827.
    e. 208,000,000.

Open-Ended Questions

1. A transcriptional map of the P element in *Drosophila* is shown on the following page. Recall that the P transposable element splices exons 0, 1, 2, and 3 in the germ-line to produce an mRNA that encodes a functional transposase, but only splices exons 0, 1, and 2 in somatic cells. You have isolated and done restriction mapping of the genomic DNA of the P element, as well as cDNAs corresponding to the germ-line and somatic P element mRNAs.

```
         exon 0     exon 1       exon 2        exon 3
         *******    *********     ********      *******

    <——————————————————————————————————————————>
```

start of   end of
transcription   transcription
(both soma and   (both soma and
germ-line)   germ-line)

    a.   You find that in the P element genomic DNA and somatic cDNA, there are three *Eco*RI restriction recognition sites, but none in the germ-line cDNA. Where do these *Eco*RI sites map within the P element?

    b.   The P element genomic DNA and somatic cDNA do not have any SalI restriction recognition sites, but the germ-line cDNA does have one such site. Explain where this site comes from.

Solution
    a.   These sites must map within the intron between exons 2 and 3, since it is only this inton which is distinct between somatic and germ-line cDNAs.
    b.   This *Sal*I site must form by a combination of the sequences from the 3' end of exon 2 and the 5' end of exon 3. Thus, the *Sal*I site would not be present in genomic DNA or somatic cDNA, but would be present in germ-line cDNAs, since these latter cDNAs uniquely splice exons 2 and 3.

2.   You are studying the *Aldox* gene that encodes the enzyme aldehyde oxidase in different species of the fruit fly, *Drosophila*. In the species *Drosophila melanogaster*, you find that *Aldox* is expressed only in the fat body (the fly's liver), whereas in the species *Drosophila virilis*, *Aldox* is only expressed in the Malpighian tubules (the fly's kidney).

In principle, there are two ways to think about these observations. One is that the enhancer element DNA sequences are different between the two species such that there is a "fat body-specific" element in *D. melanogaster* and a "Malpighian tubule-specific" element in *D. virilis*. The alternative is that we are dealing with exactly the same enhancer element sequence in both species, but that the transcription factor that binds to these elements and activates transcription of the *Aldox* gene is expressed only in the fat body in *D. melanogaster* and only in the Malpighian tubules in *D. virilis*.

You have cloned the *Aldox* genes from each of these two species. You also have in your possession a plasmid containing the *E. coli lacZ* gene under the control of a promoter that works in *Drosophila*. When it is inserted as a transgene into a *Drosophila* chromosome, this promoter is inactive unless it is near a tissue-specific enhancer element. Transgenes can be introduced into the germ-line of *D. melanogaster* and *D. virilis*.

a.   Propose a transgene-based experiment to distinguish the possibilities that it is (1) the *Aldox* enhancer elements that are different between the two species or (2) that it is the tissue-specific expression of the transcription factor that binds to the enhancer element that is different.

b.   Suppose that you find that one of the differences between the *Aldox* genes of *D. melanogaster* and *D. virilis* is that there is a chromosomal inversion that has one of its breakpoints within the transcriptional regulatory region of the *D. virilis Aldox* gene. You realize that this finding suggests that the enhancer elements of the two species might be different. Why?

Solution

a.   One obvious experiment would be to make a transgene in which the regulatory regions of the *D. virilis Aldox* gene (we will call it *Dvir/Aldox*) and the *D. melanogaster Aldox* gene (*Dmel/Aldox*) are placed in reporter gene constructs upstream of the *E. coli lacZ* gene. These reporter gene constructs would then be transformed into both *D. melanogaster* and *D. virilis*. Regulatory region fragments exhibiting proper regulation would be identified (i.e., a fragment of the *Dvir/Aldox* gene that, when transformed into *D. virilis*, expresses *lacZ* in the Malpighian tubules, or a fragment of the *Dmel/Aldox* gene that, when transformed into *D. melanogaster*, expresses *lacZ* in the fat body). These fragments would then be used in follow up reporter gene experiments.

The experiment would then be to ask what the pattern of reporter gene expression was when the hosts for the two reporter genes were swapped (*Dvir/Aldox* transgene expressed in *D. melanogaster*, and *Dmel/Aldox* expressed in *D. virilis*). If expression of the transgene reflected the species that it came from (e.g., if *Dvir/Aldox* showed *lacZ* expression in the Malpighian tubules, regardless of whether the host were *D. melanogaster* or *D. virilis*), then we would conclude that the species differences in expression were due to different enhancer elements within each gene. Alternatively, if the result were that the expression of each transgene were the same in a given host (i.e., *Dvir/Aldox* and *Dmel/Aldox* were both expressed in the Malpighian tubules in *D. virilis* hosts but were both expressed in the fat boy in *D. melanogaster* hosts), then we would conclude that the differences do not lie with the enhancer elements, but rather with the trans-acting factors (most likely, transcription factors) present in each host.

b.   Identification of a chromosomal inversion that distinguishes the two genes is reminiscent of the ability of chromosomal rearrangements to produce gain-of-function dominant mutations, such as the *Tab* inversion in *Drosophila*. Thus, a tempting possibility is that the chromosomal inversion in *D. virilis* simultaneously removes the fat body enhancer from being contiguous to the *Aldox* gene, and at the same time fuses a Malpighian tubule enhancer to *Aldox*.

3. A normal immunoglobulin heavy chain gene consists of 200 V segments, 20 D segments, 5 J segments and 1 C segment.

    a. How many possible immunoglobulin heavy chains can be produced by rearrangement among all of these segments?

    b. Suppose that an individual is heterozygous for two deletion-bearing immunoglobulin heavy chain genes. One gene is deleted for 15 V segments and 7 D segments. The other is deleted for the constant segment. In such a heterozygous individual, how many possible immunoglobulin heavy chains can be produced by DNA rearrangement?

Solution

    a. The number of heavy-chain genes that can be produced combinatorially is $(200)(20)(5) = 20,000$.

    b. The gene that is deleted for the constant region will never make a functional immunoglobulin heavy chain gene. Hence, heavy chains will only be produced from the gene deleted for some of its V and D segments. The deletion reduces the number of combinations to $(185)(13)(5) = 12,025$.

4. In humans, the *Tfm* (Testicular feminization) gene, located on the X-chromosome, encodes the testosterone receptor. The testosterone receptor, a nuclear hormone receptor, is required in every cell that produces a different phenotype in each of the two sexes.

    a. XY individuals hemizygous for a loss-of-function *Tfm* mutation are phenotypic females. Explain the basis of this mutant phenotype.

    b. An otherwise wild-type, XXY individual is phenotypically male. In each of the somatic cells, one of the two X chromosomes is inactivated by the normal process of dosage compensation. Recall that X inactivation occurs at random early in development and that the decisions of which X is inactivated in a given cell is passed on to all progeny of that cell. If an XXY individual is heterozygous for the same loss-of-function *Tfm* mutation described in a., what would be the sexual phenotype of this individual?

Solution

    a. In the absence of a functional testosterone receptor, even though XY, *Tfm* individuals form a testes and secrete testosterone, they cannot respond to the testosterone signal. Because the testosterone receptor is a transcription factor that is necessary to activate male-specific gene transcription and repress female-specific transcription, tissues remain in the default female developmental pathway.

    b. This XXY, *Tfm*/+ individual will be a mosaic of cells, some of which have functional testosterone receptor and others of which do not. The cells with functional testosterone receptor will develop a male secondary sexual phenotype whereas those without receptor will develop as phenotypically female cells. Hence, the individual will be a mixture of phenotyically male and female tissue.

5.   You have a cDNA clone corresponding to the *Drosophila* female-specific late *Sxl* mRNA. You engineer a transgene in which this cDNA is joined to the regulatory elements and promoter of the *dsx* gene. If expressed, this *dsx/Sxl* transgene would produce a functional SXL protein. If this *dsx/Sxl* transgene were inserted into the *Drosophila* genome, what would the somatc sexual phenotype be for:

    a.   XX, *dsx/Sxl* diploid individuals?

    b.   XY, *dsx/Sxl* diploid individuals?

    c.   XX, *dsx/Sxl*, *tra/tra* diploid individuals (homozygous for a loss-of-function mutation in the *tra* gene)?.

Solution

    a.   Female. Even without this transgene, such individuals would be female.

    b.   Female. The *dsx* promoter is consitutively active, but male vs. female-specific mRNAs are regulated at the level of alternative splicing. Thus, the *dsx* promoter on the transgene will cause the female-specific *Sxl* cDNA to be expressed in all cells. (In addition, this will induce female-specific splicing of the endogenous *Sxl* gene.) The active SXL protein will promote female-specific splicing of the *tra* mRNA, which in turn will promote female-specific splicing of the *dsx* transcript. The female-specific DSX protein will be produced, which will have the effect of repressing male-specific gene expression and therby will produce a female phenotype.

    c.   Male. Even though female-specific active SXL protein will be produced by virtue of the *dsx/Sxl* transgene, the absence of functional tra protein will cause *dsx* to splice in the male-specific mode.

C H A P T E R

24

Genetics and Cellular Differentiation

Multiple-Choice Questions

1.   In the worm *C. elegans*, the vulva (the opening of the uterus to the outside world) develops through a series of paracrine interactions of the anchor cell and hypoderm cells of the vulva equivalence group (VEG). The paracrine signal from the anchor cell is a polypeptide hormone that binds to a receptor tyrosine kinase (RTK) on the vulva equivalence group cells. What fates would the VEG cells adopt if the worm were homozygous for a mutation that inactivated the protein kinase activity of the RTK?

    a.   All VEG cells would adopt a primary fate.

    b.   All VEG cells would adopt a secondary fate.

    c.   All VEG cells would adopt a tertiary fate. •

    d.   None of the normal primary, secondary, or tertiary fates would be adopted.

    e.   The two VEG cells nearest the anchor cell would adopt secondary fates and all of the other VEG cells would adopt tertiary fates.

2.   You are studying the molecular biology of hormone receptors. If a transgene were constructed in which the DNA-binding domain of the thryoid hormone receptor was replaced by the DNA-binding domain of the estrogen receptor, which of the following would occur?

    a.   The transgenic receptor would respond to thyroid hormone, and activate thyroid hormone-responsive genes.

    b.   The transgenic receptor would respond to estrogen, and activate estrogen-responsive genes.

    c.   The transgenic receptor would respond to estrogen, but activate thyroid hormone-responsive genes.

    d.   The transgenic receptor would respond to thyroid hormone, but activate estrogen-responsive genes. •

    e.   All of the above

3.   Endocrine and paracrine signaling share which of the following properties?

    a.   Secretion into the bloodstream

    b.   Act upon neighboring target cells

    c.   Activate receptors on the cell surface or within target cells •

    d.   Secretion from endocrine glands

    e.   None of the above

4.   Many signal transduction cascades involve the cyclical activity of a GDP/GTP-binding protein such as that encoded by the *ras* proto-oncogene. Which of the following mutations in *ras* might convert it to a dominant oncogene state?

    a.   a mutation that prevents ras protein from binding GTP

    b.   a mutation that blocks the hydrolysis of GTP to GDP by ras protein •

    c.   a mutation that blocks interaction of *ras* with its downstream serine/threonine kinase

    d.   a nonsense mutation in the fifth codon downstream of the translation initiation codon

    e.   None of the above

5.   Many signal transduction cascades involve the cyclical activity of a GDP/GTP-binding protein such as that encoded by the *ras* proto-oncogene. Which of the following mutations in *ras* would completely block ras-mediated signal transduction?

    a.   a mutation that prevents ras protein from binding GTP •

    b.   a mutation that blocks the hydrolysis of GTP to GDP by *ras* protein

    c.   a mutation that blocks interaction of *ras* with its downstream serine/threonine kinase •

    d.   a nonsense mutation in the fifth codon downstream of the translation initiation codon •

    e.   None of the above

6.   Which of the following processes can lead to dominant oncogene mutations?
      a.   a mutation that completely deletes the DNA sequences of the proto-oncogene
      b.   a mutation that creates a constitutive enzyme activity •
      c.   a mutation that leads to the overexpression of a normal protein •
      d.   a mutation that produces a fusion of a transcription unit of the proto-oncogene to enhancer elements of another gene •
      e.   All of the above

7.   Which of the following features of the cytoskeleton contribute to the ability to create cellular asymmetries?
      a.   motor proteins operate unidirectionally
      b.   microfilaments and microtubules are inherently asymmetric
      c.   microtubules are arranged via microtubule organizing centers
      d.   None of the above
      e.   All of the above •

8.   Cyclins act to control the progression of the cell cycle by
      a.   inhibiting the action of Rb protein.
      b.   direction interaction with microtubule proteins in building the spindle apparatus.
      c.   complexing with and activating cyclin-dependent kinases. •
      d.   None of the above
      e.   All of the above

9.   The normal activity of the retinoblastoma (Rb) protein is to
      a.   induce cyclin-CDK complex formation.
      b.   inhibit p53 activity.
      c.   complex with the transcription factor E2F, except when Rb is inactivated through its phosphorylation. •
      d.   suppress transcription of E2F.
      e.   negatively regulate progression from G1 to S of the cell cycle. •

10.   Failure to have functional Rb protein in an individual
      a.   leads to absence of the negative Rb-E2F regulation of the G1 to S transition during the cell cycle. •
      b.   blocks the initiation of anaphase of the cell cycle.
      c.   blocks progression from G1 to S of the cell cycle.
      d.   can promote development of retinoblastoma tumors. •
      e.   has no effect on the cell cycle.

11.   Which of the following observations support the idea that cancers arise through an accumulation of multiple mutations promoting cell proliferation in a single cell?
    a.   A specific translocation between chromosomes 9 and 22 (the Philadelphia chromosome) is seen in almost all patients with chronic myelogenous leukemia.
    b.   Within a single tumor, several regions of the genome display chromosomal anomalies. •
    c.   In a tumor from a female patient, all of the tumor cells have the same X chromosome inactivated.
    d.   Some tumor suppressor genes cause defects in enzymes required for proper repair of damaged DNA. •
    e.   While retinoblastoma is inherited as a dominant, on a cellular basis, it is recessive. •

12.   In the nematode worm, *C. elegans*, par (partition-defective) mutations are defective in the cytoskeletal-based production of an asymmetric cleavage furrow at the first mitotic division of the zygote.

A self-cross of *par*/+ worms produces all normal progeny, whereas a self-cross of *par*/*par* worms generates progeny in which all embryos have an abnormal (symmetric) first mitotic division. A cross of *par*/*par* female worms to +/+ males also produces progeny that uniformly have an abnormal (symmetric) first mitotic division. Which of the following explanations can account for these genetic data?
    a.   The first mitotic division phenotype is based purely on the zygotic *par* genotype.
    b.   The *par* gene acts maternally to participate in the cytoskeletal asymmetries of the first mitotic division. •
    c.   There is both a maternal and a zygotic component to the function of *par* in regulating the first mitotic division.
    d.   The *par* mutation is dominant.
    e.   None of the above.

13.   You are studying the cell cycle in the budding yeast, *Saccharomyces cerevesiae*. You find that, when grown at restrictive temperatures, a temperature-sensitive mutation in a gene encoding an enzyme required for DNA replication causes cells to arrest with very small buds. This phenotype suggests that
    a.   there is a checkpoint that prevents cells in S phase from progressing into mitosis until chromosome replication occurs.
    b.   the normal activities of DNA replication genes are an essential part of the cell cycle. •
    c.   there must also be a defect in microtubule assembly in the spindle apparatus.
    d.   DNA synthesis is dispensable for cell growth.
    e.   there is a checkpoint that prevents cells at metaphase
from entering anaphase.

14. Many tumor cells are able to proliferate uncontrollably in cell culture. Transformation of wild-type alleles of tumor suppressor genes mutated in such cells (such as Rb) can cause the cultured cells to have a reduced proliferative ability. For oncogene mutations, however, the addition of wild-type alleles does not have any effect on proliferation. Which of the following statements help explain these observations?
   a.   Tumor suppressor genes can regulate DNA repair.
   b.   Many proto-oncogenes encode elements of signal transduction pathways.
   c.   Oncogene mutations are in essence gain-of-function dominant mutations. •
   d.   Tumor suppressor mutations only promote proliferation when the cell is homozygous for a knockout of these loss-of-function mutations. •
   e.   None of the above

15. One or more characteristic gene fusions between an immunoglobulin gene cluster and a proto-oncogene are frequently detected in leukemias and lymphomas of B lymphocytes. In many of these gene fusions, the transcription unit of the proto-oncogene remains intact. In such cases, the oncogenic mutation is due to
   a.   lack of transcription of the proto-oncogene in B lymphocytes.
   b.   misregulation of the proto-oncogene transcription unit by the immunoglobulin gene enhancer element. •
   c.   misexpression of the proto-oncogene in B lymphocytes. •
   d.   rearrangement of the proto-oncogene as part of immunoglobulin gene rearrangement.
   e.   None of the above

Open-Ended Questions

1.   The wild-type Rb (retinoblastoma) protein functions to sequester the transcription factor E2F in the cytoplasm. At the appropriate time in the cell cycle (G1), a cyclin-dependent protein kinase/cyclin complex phosphorylates Rb, which then releases E2F so that it can act as a functional transcription factor for genes involved in S phase of mitosis.
   a.   In cells of the retina homozygous for a loss-of-function Rb mutation, do you expect E2F to be sequestered in the cytoplasm or in the nucleus? Why?
   b.   Why do such loss-of-function Rb mutations contribute to tumor growth?
   c.   If a cell were doubly homozygous for loss-of-function mutations in the Rb and E2F genes, would you expect those mutations to promote tumor growth? Why or why not?
   d.   What would you expect to be in terms of the effect on the cell cycle of a mutation in Rb that irreversibly binds E2F, even when Rb is phosphorylated?

Solution
   a.   Nucleus. R2b's wild-type function is to sequester E2F in the cytoplasm. In the absence of R2b, E2F is free to be transported into the nucleus.

b.   E2F acts to transcribe genes encoding enzymes involved in DNA synthesis. This allows the G1 to S transition to occur, thereby contributing to unregulated cell proliferation.

c.   These doubly mutant cells would not promote tumor growth. Since there is no active E2F, there would be no G1 to S transition and hence cell proliferation would not be promoted.

d.   A mutation of Rb that irreversibly binds E2F would constitutively sequester E2F in the cytoplasm and hence would block progression of the cell cycle.

2.   Vulva development in the nematode *C. elegans* involves the anchor cell and hypodermal cells of the vulva equivalent group (VEG). Loss-of-function mutations in the gene lin-3 and let-23 fail to produce a vulva and all VEG cells adopt the tertiary fate. The lin-3 gene encodes an EGF-like polypeptide ligand, and let-23 encodes the receptor tyrosine kinase (RTK) that is activated by this ligand.

a.   What do these observations tell us about the nature of anchor cell-VEG cell interactions?

b.   How could the analysis of genetic mosaics (animals in which cells are mixtures of two different genotypes) be used to analyze the roles of lin-3 and let-23 in the anchor cell and the VEG cells.

c.   What phenotype would you expect of a mutation that produced a constitutively active let-23 RTK?

Solution

a.   These observations suggest that signaling between the anchor cell and the VEG cells occurs by RTK-mediated intercellular communication. The phenotypes of the mutant individuals suggest that the tertiary fate is the default in the absence of the activation of the RTK, and that the primary and secondary fates depend upon the presence of this signaling pathway.

b.   The observations described in a. do not tell us how the ligand and the RTK contribute to communication between the anchor cell and the VEG cells. Does the ligand get secreted from the anchor cell and act upon the VEG cells to generate the primary and/or secondary fates? Alternatively, is this a signal back from the VEG cells to the anchor cell as part of the communication between these cell types?

Mosaic analysis would allow us to determine which cells require lin-3 or let-23 in order to generate a normal phenotype. For example, we could produce a mosaic animal in which the anchor cell was wild-type for lin-3 and the VEG cells were mutant. If all such animals had a vulva and produced normal primary and secondary cells, we would then conclude that lin-3 function was unnecessary in the anchor cell for normal vulva development. (In reality, such mosaic experiments reveal that lin-3 activity is required in the anchor cell and let-23 in the primary and secondary VEG cells.)

c.   Because RTK activity is required to shift cells from the tertiary to primary or secondary fates, we would expect that a constitutively active RTK would produce a worm with multiple vulvae, in which all VEG cells have either a primary or secondary fate, and that no VEG cells would remain in the tertiary pathway.

3.   You are studying the genetic basis of cancer. You discover that in patients with a particular type of leukemia, their leukemic B lymphocytes display a translocation between one of the immunoglobulin genes and a gene encoding a cytoplasmic protein kinase.
   a.   Suggest a basis for the tumorigenic activity of this translocation.
   b.   In one patient, some of the cells containing the translocation have lost their capacity for rapid proliferation. Suggest a type of mutation in the cytoplasmic protein kinase that might have led to this loss of proliferative ability.

Solution
   a.   One likely possibility is that the protein kinase transcription unit has been fused to the enhancer elements of the immunoglobulin gene, such that the protein kinase is now misexpressed in B lymphocytes.
   b.   Any loss-of-function mutation that inactivates the protein kinase gene on the translocated chromosome would have the effect of nullifying the misexpression of the protein kinase in B lymphocytes.

4.   Retroviruses that carry viral oncogenes such as *v-erbB* or *v-ras*, and retroviruses such as MMTV represent two different themes of how oncogenic mutations are introduced into the genome. Describe how these two kinds of retroviruses contribute to tumor production.

Solution
All of these viruses contribute dominant oncogene mutations to the genome. In addition, as part of the process of infection, these retroviruses reverse transcribe DNA copies of their RNA genomes and insert these DNA copies at more or less random sites in the genome. However, *v-erbB* and *v-ras* contribute to tumorigenesis through the production of altered proteins whereas MMTV causes the expression of a structurally normal protein in a tissue in which it is not normally expressed.

Viral oncogenes such as *v-erbB* or *v-ras* have altered polypeptide sequences relative to their wild-type progenitors (proto-oncogenes). The altered proteins allow signal transduction pathways that promote growth to be activated in an unregulated fashion. For example, v-ras protein fails to hydrolze GTP to GDP and thereby remains bound to GTP. In this state, it continuously activates its downstream serine/threonine kinase. In contrast, the wild-type *ras* proto-oncogene only binds GTP transiently and in response to stimulation by the Sos protein, which in turn is activated by growth factor binding to a receptor tyrosine kinase. In a similar vein, *v-erbB* encodes a mutated form of the EGF receptor, in which signaling is constitutive and is independent of binding to the EGF ligand.

The MMTV retrovirus causes genes to be misexpressed in mammary gland tissue. The MMTV carries a special transcriptional enhancer element called an HRE (hormone response element) which is only activated by steroid hormone receptors found in the mammary gland of mice. Many insertions of the MMTV have no oncogenic effect. However, by chance, the retrovirus occasionally inserts near a gene that is not normally expressed in mammary glands and that functions, when expressed, to promote cell proliferation. In such cases, the HRE activates transcription of the gene in mammary gland tissue, and the product of this gene then promotes cell proliferation and tumorigenesis.

5.   You are studying signal transduction pathways in *Drosophila* and know that the genetic anlaysis of genotypes simultaneously carrying mutations in two genes is extremely helpful in elucidating biochemical or cellular pathways of gene action.

You have the following *Drosophila* mutant strains in a receptor tyrosine kinase (RTK) pathway necessary for development of the photoreceptor cells (cell R7) in the *Drosophila* eye:

Mutant (1): a nonsense mutant in the gene encoding the secreted ligand for the RTK. Homozygotes for this mutant completely lack R7 cells.

Mutant (2): a constitutively active mutation in the gene encoding the RTK itself. Heterozygotes or homozygotes for this mutation make extra R7 cells.
>   a.   What would be the phenotype of the (1) (2) double mutant homozygote?
>   b.   You also have mutant strain (3), which is a mutation in the gene encoding the ras protein for this RTK pathway and that causes the ras protein to not hydrolyze GTP to GDP. What R7 phenotype do you expect for homozygotes for this mutation?

Solution
>   a.   The (1) (2) double homozygote should have extra R7 cells, since the activity of the RTK mutation should be constitutive and should be independent of the presence of the secreted ligand.
>   b.   Extra R7 cells. This type of ras mutation should produce a constitutively signaling ras protein. Since the constitutive phenotype for the RTK is the production of extra R7 cells, we would expect the same to be true for the ras mutation.

C H A P T E R

25

Developmental Genetics: Cell Fate and Pattern Formation

Multiple-Choice Questions

1. Targeted gene knockouts in mice
   a. involve transplantation of mutated ES cells into host blastocysts. •
   b. are produced by introduction of foreign DNA into ES cells in culture. •
   c. involve mutagenesis of ES cells with X-rays.
   d. require identification of ES cells carrying the appropriately mutated gene. •
   e. All of the above

2. In early *Drosophila* development, the pattern of gap gene expression
   a. is dependent upon previous expression of the segment polarity genes.
   b. is determined by the activity of the anterior-posterior morphogens BCD and HB-M. •
   c. determines the pattern of expression of the pair-rule genes. •
   d. determines the pattern of expression of the homeotic genes. •
   e. contributes to dorsal-ventral patterning of the embryo.

3.   In *Drosophila*, the *nos* gene contributes to anterior-posterior patterning by inhibiting translation of HB-M in the posterior portion of the oocyte. What would happen if the 3' untranslated region (UT) of the HB-M mRNA were replaced with the 3' UT of a gene not sensitive to translation regulation?
   a.   There would be a loss of posterior segments. •
   b.   There would be no change in the segmentation pattern of the embryo, since the BCD gradient would remain unchanged.
   c.   There would be a loss of anterior segments.
   d.   Normal dorsal-ventral pattern formation would be disrupted.
   e.   HB-M protein would be produced at high levels everywhere in the embryo. •

4.   Which of the following are characteristics of dominant maternal effect mutations?
   a.   Mutant offspring arise from heterozygous mutant mothers. •
   b.   Wild-type offspring arise from crosses of heterozygous mutant fathers, regardless of the genotype of the mothers.
   c.   Heterozygous mutant fathers crossed to homozygous normal mothers produce normal offspring. •
   d.   Mutant offspring arise from homozygous mutant fathers crossed to normal mothers.
   e.   When mutant offspring arise, they account for 50% of the offspring.

5.   In early *Drosophila* development, *gap* gene expression is always normal in embryos homozygous for pair-rule mutations, whereas pair-rule expression is always mutant in gap gene mutant genotypes. From these observations, we conclude that
   a.   pair-rule and *gap* gene expression patterns arise independently of one another.
   b.   *gap* gene expression is determined by the distribution of anterior-posterior morphgens.
   c.   proper pair-rule gene expression is dependent upon previous *gap* gene expression. •
   d.   *gap* genes encode transcription factors.
   e.   *gap* gene expression is established prior to, and independent of pair-rule gene expression. •

6.   Establishment of the dorsal-ventral axis of the *Drosophila* embryo:
   a.   involves the production of localized DL mRNA on the ventral side of the developing oocyte.
   b.   occurs via at least two paracrine signals transmitted between the developing oocyte and the surrounding somatic follicle cells. •
   c.   involves the zygotically expressed DL morphogen.
   d.   All of the above
   e.   None of the above

7.   Place the following *Drosophila* developmental stages in their chronological order, beginning with the events that occur inside the ovary of the mother

3   a.   cellular blastoderm
2   b.   fertilization
1   c.   oogenesis
4   d.   gastrulation
5   e.   larva

8.   The DL protein is the dorsal-ventral morphogen encoded by the *dl* gene. The TL protein encoded by the *Tl* gene is a transmembrane receptor. Embryos derived from homozygous dl or homozygous *Tl* mutant mothers are completely dorsalized. In wild-type cellular blastoderm-staged embryos, you find that DL protein is present in exclusively in the nucleus in ventral cells, partially in the nucleus and partially in the cytoplasm in lateral cells, and exclusively in the cytoplasm in dorsal cells. In contrast, in cellular blastoderm-staged embryos derived from homozygous mutant *Tl* mothers, DL protein is exclusively cytoplasmic in location in all cells of the embryo. Which of the following statements are valid conclusions from these observations?
   a.   The nuclearly localized DL protein is the active form of the molecule. •
   b.   TL receptor activity is required to activate DL protein on the ventral and lateral sides of the embryo. •
   c.   The cytoplasmically localized DL protein is the active form of the molecule.
   d.   The embryo appears to be subdivided into a very few domains based on the level of DL nuclear localization. •
   e.   None of the above

9.   The homeotic, *gap*, pair-rule and segment polarity genes all control
   a.   aspects of anterior-posterior pattern formation. •
   b.   the number of segments.
   c.   the developmental fates (identities) of the individual segments.
   d.   dorsal-ventral pattern formation.
   e.   the production of the anterior-posterior morphogens.

10.   The homeotic genes, in contrast to the *gap*, pair-rule and segment polarity genes, control
   a.   interactions between the follicle cells and the oocyte during oogenesis.
   b.   the number of segments.
   c.   the developmental fates (identities) of the individual segments. •
   d.   dorsal-ventral pattern formation.
   e.   the production of the anterior-posterior morphogens.

11. Homeosis refers to
     a.   the production of an animal with too many segments.
     b.   the development of an animal missing a major portion of the dorsal-ventral axis.
     c.   the production of a body part in an ectopic location, in place of the part that should normally develop there. •
     d.   None of the above
     e.   All of the above

12. How does the cytoskeleton contribute to the polarity of the *Drosophila* oocyte?
     a.   Through microtubule control of the orientation of mitotic cleavages
     b.   Through microtubule-directed placement and movement of the oocyte nucleus which in turn localizes the site of the GRK secreted signal •
     c.   Through microfilament-directed asymmetric cell divisions
     d.   All of the above
     e.   None of the above

13. The homeotic (HOM-C) genes and the Sxl gene have in common that
     a.   they are master switch genes for the developmental fate of cells.
     b.   they encode proteins that regulate mRNA splicing patterns.
     c.   they must be expressed throughout the lifetime of a cell to maintain its proper fate. •
     d.   they are both regulated by the X:A ratio of the fly.
     e.   different mechanisms control their initial expression and their maintenance. •

14. You are studying mammalian segmentation by generating zygotically acting mutations in the mouse. You identify a recessive mutation which causes all seven cervical vertebrae (the vertebrae in the neck) to be missing. Given the striking evolutionary conservation of many aspects of development in mice and flies, one way to figure out what gene is mutated is that it might be a gene affecting an analogous process in *Drosophila*. What type of segmentation gene in *Drosophila* is most analogous to the loss of seven cervical vertebrae phenotype produced by your mouse mutation?
     a.   anterior-posterior morphogen gene
     b.   *gap* gene •
     c.   pair-rule gene
     d.   segment polarity gene
     e.   homeotic gene

15.   One way that the SPZ protein was identified as the ligand for the TL receptor in dorsal-ventral patterning of the *Drosophila* embryo was by purifying SPZ protein and micro-injecting it into the perivitelline space (the space between the eggshell and the embryo itself). TL receptor is known to be expressed on all cells of the embryo. If SPZ protein remained at the site of the injection, and if the embryos you injected were derived from mothers homozygous for the *spz* gene, what would you expect the dorsal-ventral phenotype of the injected embryos to be?

    a.   dorsalized everywhere except at the injection site, where ventral and lateral tissue would be found •

    b.   dorsalized everywhere

    c.   ventralized everywhere except at the injection site, where dorsal and lateral tissue would be found

    d.   ventralized everywhere

    e.   None of the above

Open-Ended Questions

1.   a.   In embryos derived from a *nanos* loss-of-function mutant homozygous female, the posterior (abdominal) segments fail to form. Explain the basis of this phenotype in terms of the effect of *nanos* on anterior-posterior positional information.

    b.   You create a transgene that had a bicoid promoter and bicoid protein coding sequences fused to the 3′ UT of *nanos*. You create embryos from mothers carrying this transgene and which are also homozygous for a loss-of-function *nanos* mutation. What phenotype do you expect for such embryos? Justify your answer.

Solution

    a.   In order for the posterior segments to form, posterior *gap* genes must be expressed. Their expression occurs in regions in which there is little or no HB-M protein being expressed. The role of NOS protein is to prevent translation of HB-M mRNA in the posterior of the animal. *nanos* mRNA is localized at the posterior pole and upon translation, a gradient of Nanos protein forms with its highest concentration at the posterior pole. HB-M mRNA, which is uniform in concentration throughout the early embryo, is selectively translated only where Nanos protein is absent (in the anterior of the early embryo). This creates a shallow anterior to posterior gradient of HB-M protein, in which concentrations of HB-M in the anterior half of the embryo are sufficient to repress transcription of the posterior *gap* genes.

    b.   This embryo will be dicephalic, that is, it will have two anterior sets of segment in mirror image to one another. The *bicoid* mRNA will be localized to both poles, and hence Bicoid protein will be expressed in mirror image gradients. Further, HB-M mRNA will be translated everywhere, since there is no protein present. Thus, there will be no HB-M protein gradient, but rather HB-M protein will be expressed at levels characteristic of the anterior part of the embryo.

2.   In the mouse, a mutation causes one of the *Hox* genes that is normally expressed only in the lumbar (posterior) vertebral segments to be expressed in the cervical vertebrae as well.
   a.   Based on what you know about homeotic gene function in flies and mice, would you expect such a mutation to be dominant or recessive in producing its mutant phenotypes? Justify your answer.
   b.   Would you expect these mutant phenotypes to affect the number of vertebrae or the type of vertebrae that are produced or both? Justify your answer.

Solution
   a.   Dominant. Typically, the HOM-C and *Hox* genes are regulated such that, in a homozygous knockout for one of these genes, the gene that is normally expressed immediately anterior to that gene becomes activated in the normal expression domain of the knocked out gene. Thus, loss-of-function mutations cause genes to be expressed more posteriorly than normal. On the other hand, gain-of-function dominant mutations typically cause a posterior gene to be expressed in a more anterior segment. These lead to the production of a gain-of-function phenotype in which posterior segmental material is expressed anteriorly.
   b.   The type of vertebrae should be affected but not their numbers. The *Hox* genes, like the insect HOM-C genes, are segment identity genes, controlling the qualitative properties of the segments that are produced. Other genetic pathways are employed to determine segment number.

3.   Toll-Dominant mutant alleles cause the deletion of the extracellular domain of the TL transmembrane receptor. Toll-Dominant mutants cause a complete ventralization of the embryonic body plan. Loss-of-function mutations in spaetzle, which encodes the SPZ ligand for the TL receptor, cause a completely dorsalized embryonic phenotype. Mothers that are homozygous for a loss-of-function mutant allele of spaetzle and heterozygous for Toll-Dominant produce completely ventralized embryos. Explain these observations in terms of the way the dorsal-ventral axis is established in the *Drosophila* oocyte and embryo.

Solution
The TL receptor is ubiquitously expressed on cells of the oocyte. Ordinarily, it is only activated by the SPZ ligand, which in turn is only secreted in active form in the perivitelline fluid on the ventral side of the oocyte. The Toll-Dominant mutants produce a constitutively active receptor that can initiate its intracellular signal transduction cascade even in the absence SPZ ligand. The TL receptor signal then acts to cause the Dorsal (DL) protein to be phosphorylated such that it no longer is bound to Cactus protein in the cytoplasm. In this way, DL protein is able to move into the nucleus and act as a transcription factor regulating the expression of the D/V cardinal genes.

4. The cytoskeleton of the oocyte plays a major role in the production of the major body axes of the embryo. Describes three events in the production of the body axes to which the cytoskeleton contributes.

Solution

Event 1: The localized expression of gurken (*grk*) mRNA is essential for the initial establishment of the A/P and D/V axes of the follicle cells. Initially, the oocyte nucleus is found at a posterior position. Presumably because *grk* mRNA is produced by the oocyte nucleus and because it will produce a protein that will be secreted, *grk* mRNA is localized (probably in the rough endoplasmic reticulum) near the nucleus. Thus, when GRK protein is processed through the cell's secretory pathway, it will be most concentrated in the vicinity of the nucleus. When the nucleus is posterior, the secreted GRK protein acts upon the EGF receptor tyrosine kinase of the nearest follicle cells (the polar follicle cells) and the signaling pathway that is activated causes those cells to adopt a posterior polar cell fate. As part of this interaction, the microtubules, which initially are in an MTOC configuration, reorganize and polarize such that the plus end is posterior and the minus end is anterior. During this process, the nucleus is moved to an anterior position near one side of the oocyte. The GRK signal is then produced more strongly on this side of the oocyte, causing the neighboring follicle cells to adopt a dorsal fate. Through these two sequential signaling events with one ligand, the follicle cells at the posterior pole of the oocyte become distinct from the anterior cells, and the dorsal follicle cells become distinct from the ventral.

Event 2: The localized expression of bicoid (*bcd*) mRNA at the anterior end of the oocyte is necessary to establish the BCD morphogen gradient. While the exact mechanism is not understood, it is very clear that *bcd* mRNA becomes localized in a cap at the anterior end of the developing oocyte, probably through interactions of its 3' untranslated region with proteins that bind it to one end of the polarized microtubules in the oocyte. This localized mRNA is then responsible for producing the gradient of BCD protein, one of the two A/P morphogens.

Event 3: The localized expression of *nanos* (*nos*) mRNA at the posterior pole of the oocyte is necessary to establish the NOS protein gradient. Through its 3' untranslated region, *nos* mRNA becomes localized to the posterior pole of the oocyte, presumably through a complex set of interactions that indirectly anchor the mRNA to the other end of the polarized microtubules. This localized mRNA is then responsible for producing the gradient of NOS protein. NOS inhibits the translation of HB-M mRNA, thereby creating the gradient of HB-M protein, the other A/P morphogen.

5. In *Drosophila*, mutations in the gap genes disrupt expression of one or two of the stripes of expression of each of the pair-rule genes. Mutations in some pair-rule genes, such as even-skipped (*eve*) disrupt the pattern of stripes of expression of other pair-rule genes such as fushi tarazu (*ftz*). However, mutations in *ftz* do not disrupt the pattern of pair-rule expression of the eve stripes. When the regulatory elements of the *eve* gene are

analyzed by means of a series of reporter gene constructs in which small pieces of the *eve* genomic DNA are used to drive expression of *lacZ* in transgenes, it is found that different pieces of eve regulatory DNA contain enhancers for different stripes. Thus, there appear to be different enhancers for each of the 7 pair-rule *eve* stripes. In contrast, for other pair-rule genes such as ftz, there appears to be only one enhancer that is responsible for the 7-stripe pattern (the zebra enhancer element). Explain these observations in terms of how *gap* genes and pair-rule genes form a regulatory hierarchy.

Solution
The *gap* genes are each expressed in a localized domain of the embryo and each controls the activation of pair-rule genes within its domain. The different properties of pair-rule genes such as *eve* and *ftz* show that there is a hierarchical arrangement within the pair-rule gene class as well. The direct effect of the *gap* genes in forming the correct number of segments is to activate a specific stripe of expression of pair-rule genes of the *eve* class according to the concentration of the protein encoded by that *gap* gene. In aggregate, then, all of the *gap* gene products determine the domains within which the *eve* class of pair-rule genes are activated. The pattern of activation of the *eve* class of pair-rule genes is then responsible for binding to the zebra-type enhancers of the pair-rule genes of the *ftz* class, thereby activating them in their pair-rule stripe pattern. Thus, the asymmetric expression of the *gap* genes is employed to produce a repetitive pattern of expression of the *eve* class of pair-rule genes, and in turn this repetitive pattern is employed to produce the repetitive patterns of the other pair-rule genes.

C H A P T E R

26

Population Genetics

Multiple-Choice Questions

1.   In a human population the genotype frequencies at one locus are 0.5 *AA*, 0.4 *Aa* and 0.1 *aa*. The frequency of the *A* allele is
    a.   0.32.
    b.   0.75.
    c.   0.70. •
    d.   0.20.
    e.   0.9.

2.   In a population of mice an antigen locus has two alleles *A1* and *A2*. The genotype frequencies are 0.21 *A1 A1*, 0.30 *A1 A2*, and 0.49 *A2 A2*. The frequency of the *A2* allele is
    a.   0.7.
    b.   0.46.
    c.   0.79.
    d.   0.25.
    e.   0.64. •

3.   In a *Drosophila* population there are three isozyme alleles I^1, I^2 and I^3, at one locus. The frequency of I^1 is 0.4, and the frequency of I^3 is 0.15. What is the frequency of I^2?
    a.   0.6
    b.   0.45 •
    c.   1.05
    d.   0.2
    e.   1.55

4.   An RFLP probe shows four different morphs ('alleles') in one population of insects. How many different genotypes will there be in the population in regard to this RFLP?

  a.   6 •
  b.   4
  c.   3
  d.   16
  e.   24

5.   In one plant population there are two electrophoretic alleles of the esterase gene. A survey showed the following numbers, as shown underneath the gel showing representative banding patterns.

Well->

—   —

—
—   —

        38   50   12

The frequency of the fast allele is

  a.   0%.
  b.   12%.
  c.   38%.
  d.   37%. •
  e.   3.5%.

6.   If two populations of birds with frequencies of an allele *a* of 0.2 and 0.3 are forced together by unusually high winds. The new frequency of the *a* allele in the combined population will be

  a.   0.3.
  b.   0.2.
  c.   0.5.
  d.   0.25. •
  e.   0.15.

7.   Three different human populations have the following *MN* blood group frequencies

	MM	*MN*	*NN*
Pop 1	0.2	0.2	0.6
Pop 2	0.64	0.32	0.04
Pop 3	0.81	0.18	0.01

The populations in Hardy-Weinberg equilibrium are
    a.   1 only.
    b.   2 only.
    c.   3 only.
    d.   1 and 2.
    e.   2 and 3. •

8.   In populations in Hardy-Weinberg equilibrium, the heterozygote frequency is maximal when the *A* allele equals
    a.   0.
    b.   0.2.
    c.   0.5. •
    d.   0.8.
    e.   1.0.

9.   In a tropical human population in Hardy-Wenberg equilibrium for an autosomal locus determining presence/absence of pigment in the skin, the frequency of albinism (*aa*) is 1 in 10,000. The frequency of heterozygotes is approximately
    a.   1 in 50. •
    b.   1 in 100.
    c.   1 in 1000.
    d.   1 in 75.
    e.   1 in 25.

10.   In a population in Hardy-Weinberg equilibrium, what will be the proportion of marriages between homozygotes?
    a. $p^2 + 2pq$
    b. p^4
    c. $p^4 + q^4$
    d. $p^4 + q^4 + 2p^2q^2$ •
    e. $4p^2q^2$

11.   In a population in Hardy-Weiberg equilibrium for the *A/a* locus, the frequency the *a* allele is 0.1. If the fitness of the *aa* genotype suddenly becomes 0, how many generations will it take before the frequency of the *a* allele is 0.05 ?
    a.   1
    b.   2
    c.   5
    d.   10 •
    e.   100

12.   In an experimental population of fish, the genotype frequencies are 0.4 *BB*, 0.4 *Bb*, and 0.2 *bb*. If all the *BB* fish are removed from the population, what will be the frequency of the *B* allele in the next generation?
    a.   20%
    b.   33% •
    c.   60%
    d.   40%
    e.   30%

13.   In a population in mutation/selection equilibrium for the *D/d* locus, there is selection against *dd* and only forward mutation *D -> d* If the frequency of *dd* genotypes is 4×10^{-3}, and the selection coefficient against *dd* is 2×10^{-2}, what is the mutation rate to *d*?
    a. 2×10^{-3}
    b. 4×10^{-6}
    c. 2.4×10^{-2}
    d. 5×10^{-7}
    e. 8×10^{-5} •

14.   In a population of snails, one selective force is acting against *AA* giving it a fitness of 0.6, and a different selective agent is acting against *aa* giving it a fitness of 0.7 in relation to *Aa* whose fitness can be 1.0. When these two forces reach equilibrium, what will be the equilibrium frequency of the allele *A*?
    a.   2/3
    b.   3/4
    c.   3/7 •
    d.   4/9
    e.   11/13

15.   In a population of rats, the fitnesses and frequencies of genotypes at present are

	Frequency	Fitness
HH	0.4	1
Hh	0.5	0.8
hh	0.1	0.6

What will be the frequency of the *h* allele in the next generation?
    a.   0.3 •
    b.   0.4
    c.   0.5
    d.   0.6
    e.   0.7

Open-Ended Questions

1.   Tay-Sachs disease is inherited as an autosomal recessive. In a certain large eastern European population the frequency of Tay-Sachs disease is 1%.
     a.   If the population is assumed to be in Hardy-Weinberg equilibrium, what is the frequency of the allele that causes Tay-Sachs disease?
     b.   What would be the frequency of heterozygotes?
     c.   What is the probability of two heterozygotes marrying?
     d.   In children of such marriages what would be the frequency of Tay-Sachs disease?
     e.   What proportion of all Tay-Sachs births are contributed by such marriages?

Solution
     a. q = square root of 0.01 = 0.1.
     b. $2 \times 0.9 \times 0.1 = 0.18$.

     c. $0.18^2 = 0.032$.
     d.   0.25.
     e. $0.032 \times 0.25 = 0.0081$ which is 81% of the total Tay-Sachs cases (0.0100).

2.   All humans are of either blood group *M*, *N*, or *MN*, and these are determined by two alleles: $I^M I^M$ = group *M*, $I^N I^N$ = group *N*, and $I^M I^N$ = group *MN*. In one Canadian Inuit population there were
     320 people of type *M*,
     110 people of type *N*, and
     640 people of type *MN*.

     a.   What are the frequencies of alleles I^M and I^N?
     b.   Is the population in Hardy-Weinberg equilibrium?
     c.   If the population were to begin mating randomly, what blood group frequencies would be observed in the next generation?

Solution
     a.   freq (*M*) = $[320 + (0.5 \times 640)] / 1070 = 0.6$.

     freq (*N*) = $[110 + (0.5 \times 640)] / 1070 = 0.4$.

     b.   No; if so *MM* would be $0.6^2 = 0.36$, and *nn* would be $0.4^2 = 0.16$, which are both different from the proportions observed.
     c.   Instant HWE; 0.36 of *MM*, 0.48 of *MN*, 0.16 of *NN*.

3.   Red hair (autosomal recessive) is found in approximately 4% of people in Norway. If we assume that the Norwegian population is in Hardy-Weinberg equilibrium,
     a.   What are the frequencies of the red hair (*r*) and the non-red hair (*R*) alleles?
     b.   What is the frequency of heterozygotes?
     c.   What proportion of all marriages stand a chance of having a child with red hair?

Solution

    a. q = square root of 0.04 = 0.2, so p = 0.8.

    b. $2 \times 0.2 \times 0.8 = 0.32$.

    c. $Rr \times Rr = 0.32 \times 0.32 = 0.1024$.

        $Rr \times rr = 0.32 \times 0.04 \times 2 = 0.0256$.

        $rr \times rr = 0.04 \times 0.04 = 0.0016$.

        total = 0.1296 or 12.96%.

4.   A population of insects is in Hardy-Weinberg equilibrium for a gene with alleles *A* = orange and *a* = yellow eyes. There are 91% orange and 9% yellow individuals in the population. If the fitness of the yellow phenotype suddenly drops to zero, what will be the allele frequency in the next generation?

Solution

The genotype frequencies and their contributions to the next generation can be calculated as laid out below:

		Gametes for next generation	
		A	*a*
AA	0.49	0.49	0
Aa	0.42	0.21	0.21
aa	0.09	0	0
	Totals	0.70	0.21

Therefore   new *p* will be 0.7/0.91 = 0.77

               new *q* will be 0.21/0.91 = 0 23

5.   An insect trap was set out in August and 1000 individuals from one species of mosquito were assayed for the enzyme phosphoglucomutase (PGM), by running proteins on a gel and staining for the enzyme. The results are shown below, with the numbers of individuals in each class.

Type 1	Type 2	Type 3
	—	—
		—
	—	—
416	82	502

    a.   Explain the three different staining patterns on the gel.

    b.   Calculate the frequencies of the PGM alleles.

    c.   If all individuals of genotype 2 could be eliminated, what would be the allele frequencies in the next generation?

Solution

    a.   PGM is probably a dimeric protein; type 1 is homozygous for a slow allele, type 2 is homozygous for a fast allele, and type 3 is a heterozygote showing fast, slow, and a hybrid band.

    b.   Fast: $416 + 1/2 \times 502$ out of $1000 = 0.667$.

           Slow: $82 + 1/2 \times 502$ out of $1000 = 0.333$.

    c.   See question 4 for calculation method.

           new freq $F = 251/918 = 27\%$.

           new freq $S = (416 + 251)/918 = 73\%$.

6.   A totalitarian regime vows to eliminate recessive genetic disease alleles from its population by outlawing reproduction of recessive homozygotes. It will begin with the autosomal recessive disease cystic fibrosis. In their country the cystic fibrosis allele has a frequency of 0.02 The government's goal will be to halve this frequency. How many generations will this take?

Solution

Study of the formula in the chapter reveals that if an allele frequency is expressed in the form *1/a*, then every generation of maximal selection against the recessive phenotype will add 1 to the denominator, that is, *1/a -> 1/(a+1) -> 1/(a+2)* etc. In this example the initial allele frequency is $0.02 = 1/50$. Therefore to get from 1/50 to 1/100 will take 100 - 50 = 50 human generations, or approximately 1000 years!

7.   Human fitness estimates can be used to measure mutation rates. It is estimated that the fitness of a certain autosomal recessive nerve disease is 0.4. The frequency of the disease in the population is 1 in 10,000. If reversion is ignored, what is the mutation rate to recessive allele?

Solution

The selection coefficient $s = 1-$ fitness $= 1 - 0.4 = 0.6$. Furthermore, $q^2 = 0.0001$. Rearranging the formula $q^2 = u/s$, we find that $u = q^2 s$ which $= 0.0001 \times 0.6 = 6 \times 10^{-5}$.

27

Quantitative Genetics

Multiple-Choice Questions

1. Four polygenic loci each with two alleles affect the quantitative phenotype internode length in the plant *Mimulus guttatus*. The number of different distinct polygene genotypes is
    a.   4.
    b.   12.
    c.   16.
    d.   64.
    e.   81. •

2. Four polygenic loci each with two alleles affect the quantitative phenotype internode length in the plant *Mimulus guttatus*. If each allele has the same additive effect, the number of different classes of internode length will be
    a.   4.
    b.   9. •
    c.   12.
    d.   16.
    e.   81.

3. The formula $(\Sigma\, x_i)/N$ defines

    a.   mode.
    b.   mean. •
    c.   standard deviation.
    d.   variance.
    e.   covariance.

4.   The formula $[\Sigma\,(x_i-x)^2]/N$ defines
     a.   mode.
     b.   mean.
     c.   standard deviation.
     d.   variance. •
     e.   covariance.

5.   The formula $[\Sigma\,(x_i-x)(y_i-y)]/N$ defines
     a.   mode.
     b.   mean.
     c.   standard deviation.
     d.   variance.
     e.   covariance. •

6.   Variation around the central class of a distribution is measured by
     a.   mean.
     b.   mode.
     c.   variance. •
     d.   correlation coefficient.
     e.   heritability.

7.   In a litter of five baby mice the body weights were 2, 4, 6, 8, and 10 cg. The variance of these measurements is
     a.   8. •
     b.   6.
     c.   40.
     d.   2.
     e.   5.

8.   In wheat the shade of redness in seeds is determined by polygenes. A plant with medium red seeds was selfed and gave distinct progeny classes that were dark red, red, medium red, light red and white in decreasing order of redness. The number of heterozygous polygene loci in the original plant was
     a.   2. •
     b.   3.
     c.   4.
     d.   5.
     e.   6.

9.   The function that transforms an environmental distribution into a phenotypic distribution is the
   a.   variance.
   b.   norm of reaction. •
   c.   familiality.
   d.   broad sense heritability.
   e.   narrow sense heritability.

10.   In a nematode population, the variance of length was found to be 32, and environmental variance was 8. The broad sense heritability must be
   a.   5%.
   b.   10%.
   c.   20%.
   d.   50%.
   e.   75%. •

11.   Two pure lines of plants had mean heights of 26 and 42 cm. The F_1 had a mean of 34 cm and a variance of 20. The F_2 also had a mean of 34 but a variance of 60. The broad sense heritability must be
   a.   13%.
   b.   66.7%. •
   c.   50%.
   d.   34%.
   e.   80%.

12.   In rats measurements were made of the time taken to run through a maze. The correlation between full siblings was 0.3 and between half siblings was 0.2. The broad sense heritability of maze running time in this population is
   a.   0.20.
   b.   0.80.
   c.   0.06.
   d.   0.40. •
   e.   0.50.

13.   A plant was found to be heterozygous for 16 RFLP loci. If this plant is selfed, then one of its progeny is selfed, then one of that plant's progeny is selfed, on average how many heterozygous RFLP loci will individuals in the next generation have?
   a.   2. •
   b.   3.
   c.   13.
   d.   64.
   e.   128.

14. A population of beetles had a mean weight of 30g. Individuals of 12g were selected and allowed to interbreed, and their progeny had a mean weight of 21g. The narrow sense heritability is

    a.   20%.
    b.   18%.
    c.   26%.
    d.   50%. •
    e.   80%.

15. Narrow sense heritability is a quantification of the proportion of total variance due to

    a.   phenotypic variance.
    b.   genetic variance.
    c.   additive genetic variance. •
    d.   dominance variance.
    e.   environmental variance.

16. In a chicken population the following measurements refer to fat content
$s_g^2 = 120, s_e^2 = 80$ and $s_d^2 = 38$.
The narrow sense heritability is

    a.   0.74.
    b.   0.52.
    c.   0.90.
    d.   0.41. •
    e.   0.24.

Open-Ended Questions

1. A breeding experiment was performed on pure lines of tomatoes differing in fruit weight. The plants were all grown in a controlled environment chamber providing equal environmental conditions for all individuals as far as humanly possible. The results were as follows:

Parental lines   *A* 24 g   ×   *B* 32 g

F_1 28 g

F_2	Weight (g)	Number of plants
	36	2
	34	14
	32	60
	30	108
	28	140
	26	114
	24	52
	22	18
	20	2

Show how a polygene model can explain these results, and give

    a.   a general statement of your model and

    b.   genotypes of the parents and F_1, and the polygenic 'dose' genotypes of the nine F_2 classes under your model.

    c.   Explain the frequencies of the 20 g and the 36 g F_2 classes

Solution

    a.   four polygenic loci, giving 0 to 8 polygene 'doses.' From a baseline of 20, each polygene adds one 2g dose.

    b.   *AA bb cc dd* (2 doses) × *aa BB CC DD* (6 doses) -> *Aa Bb Cc Dd* (4 doses) -> F_2

ranges from *aa bb cc dd* (0 doses) to *AA BB CC DD* (8 doses).

    c.   These are each expected at a frequency of $(1/4)^4 = 1/256$, approximately the frequencies observed.

2.   If an individual plant is heterozygous for six polygenic loci that affect height, and this plant is selfed,

    a.   how many height classes will there be in the progeny?

    b.   what will be the frequency of the tallest class of progeny?

Solution

    a.   There can be a range from 0 to 12 polygenes, a total of 13 phenotypic classes.

    b. $(1/4)^6 = 1/4096$.

3.   In a natural population of outbreeding annual plants, the variance of the total number of seeds per plant was 16. From the natural population 20 plants were taken into the laboratory and each was selfed for 10 generations. The average variance in the tenth generation in each of the 20 sets was about equal, and averaged 5.8. Estimate the broad sense heritability in this population from these data.

Solution

Repeated selfing results in homozygosity so plants within each of the 20 sets were genetically identical. Therefore the variance must have been envirinmental variance and this can be subtracted from the population variance in order to obtain genetic variance.

$$H_2 = (16 - 5.8)/16 = 10.8/16 = 68\%$$

4.   Two pure lines of *Nicotianum* (tobacco) had significantly different corolla lengths. In line 1 the average length was 30 mm, ranging from 25 mm to 35 mm with a variance of 10, in line 2 the average corolla length was 60 mm ranging from 45 to 65 and variance was 12. The F_1 had a corolla length of 45 mm ranging from 40 to 50, and the variance was 9. Upon selfing, an F_2 was produced which had an average flower length of 45mm but ranged from 10mm to 80mm, with a variance of 55.

    a.   Provide a general explanation of these results.

    b.   Calculate broad sense heritability

Solution

    a.   Polygenes can explain the general results. There is no segregation in the P or F_1 generations, but in the F_2 the heterozygous polygenes of the F_1 have segregated to give the complete range of genotypes. For example

$$AA\ BB\ CC\ dd\ ee\ ff \times aa\ bb\ cc\ DD\ EE\ FF$$
$$F_1\ Aa\ Bb\ Cc\ Dd\ Ee\ Ff$$
$$F_2\ \text{ranges from}\ aa\ bb\ cc\ dd\ ee\ ff\ \text{to}\ AA\ BB\ CC\ DD\ EE\ FF$$

    b.   The variance of the P and F_1 must be all environmental, and these are all approximately equal at around 10. The F_2 variance must be a combination of genetic and environmental variance. Therefore

$$H^2 = 55 - 10/55 = 45/55 = 82\%.$$

5.   In a population of beetles the total variance of body weight was 130. It was estimated that the environmental variance was 35, and dominance genetic variance was 45. From these data calculate heritability in the narrow sense.

Solution

Additive genetic variance must be 130 - 45 - 3 5 = 50. So $h^2 = 50/130 = 39\%$.

6.   In a large flock of turkeys body weight was 10 pounds. In a test to determine if the average weight of the flock could be increased by selective breeding, birds of weight 14 pounds were removed and allowed to interbreed. Their progeny had a body weight of 11 pounds. From this information estimate narrow sense heritability.

Solution

The selection differential was 14 - 10 = 4, and the gain was 11 - 10 = 1, so $h^2 = 1/4 = 25\%$

7. In common wheat, *Triticum aestivum*, kernel color is determined by polygenes, such that any number of *R* genes will give red, and the lack of *R* genes will give white phenotype. In one cross between a red pure line and a white pure line, the F_2 was 63/64 red and 1/64 white.
   a. How many polygenic loci are segregating in this system?
   b. Show genotypes of parents, F_1 and F_2.
   c. Different F_2 plants are backcrossed to the white parent. Give examples of polygene genotypes that would give the following ratios in such backcrosses, i. 1 red : 1 white, ii. 3 red : 1 white, iii. 7 red : 1 white.
   d. What is the formula that generally relates the number of segregating polygenic loci to the proportion of red individuals in the F_2 in such systems?

Solution
   a. 3, because $(1/4)^3 = 1/64$
   b. $A_1A_1\ A_2A_2\ A_3A_3 \times a_1a_1\ a_2a_2\ a_3a_3$

   F_1 $A_1a_1\ A_2a_2\ A_3a_3$
   F_2 $A\text{-}\ \text{---} = $ red
         aa aa aa = white
   c. *Aa aa aa* × *aa aa aa*

   Aa Aa aa × *aa aa aa*

   Aa Aa Aa × *aa aa aa*

   d. Proportion of red individuals = $1 - (1/4)^n$

8. The seed color of wheat shows a continuous distribution from white to dark red. However, in the progeny of one selfed plant there were seven distinct classes, one of white and six shades of red up to dark red. Both white and dark red were at a frequency of 1/64.
   a. Propose an explanation of these results. State your assumptions clearly.
   b. What would be the most frequent phenotypic class, and what type of genotype would it be?
   c. If a plant from the class one shade lighter than the darkest red is selfed, what will be the phenotypic proportions in the progeny?

Solution
   a.

RR RR RR = darkest shade		6 R doses, freq $(1/4)^3$
RR RR Rr	5	
RR RR rr, RR Rr RR	4	
RR Rr rr etc.	3	
RR rr rr, Rr Rr rr etc.	2	
Rr rr rr etc.	1	
rr rr rr = lightest shade	0	freq $(1/4)^3$

b.   The 3-dose class, whose color will be the third lightest class of red.

c.   *RR RR Rr* selfed will give   1/4 *RR RR RR* 6 doses
                                            1/2 *RR RR Rr* 5
                                            1/4 *RR RR rr* 4

Transparency Masters

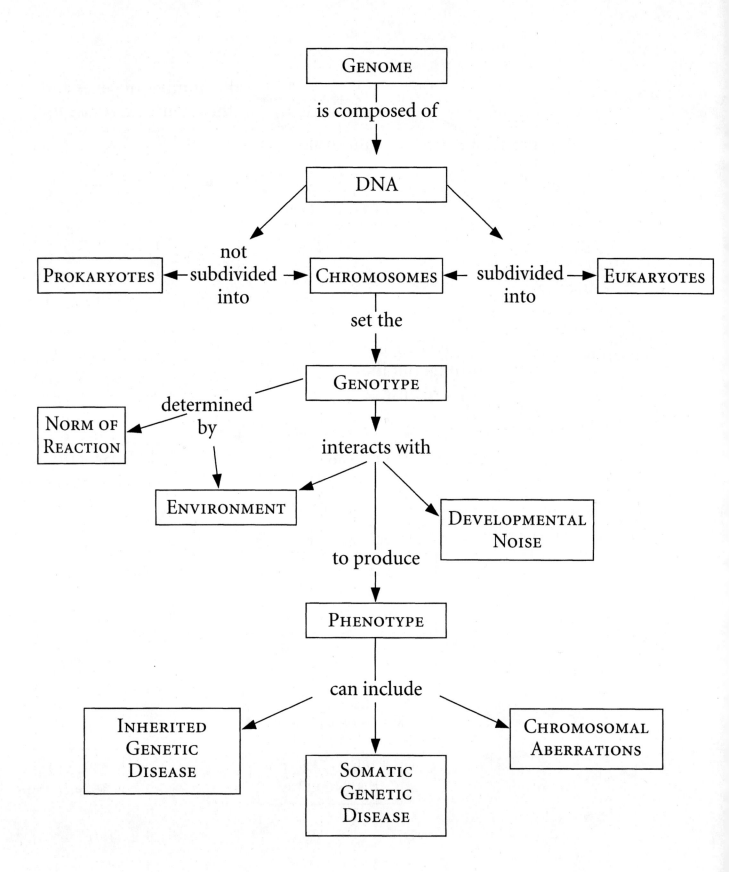

Concept Map 1-1, page 2 in Lavett: *Student Companion with Complete Solutions for IGA,* 6e

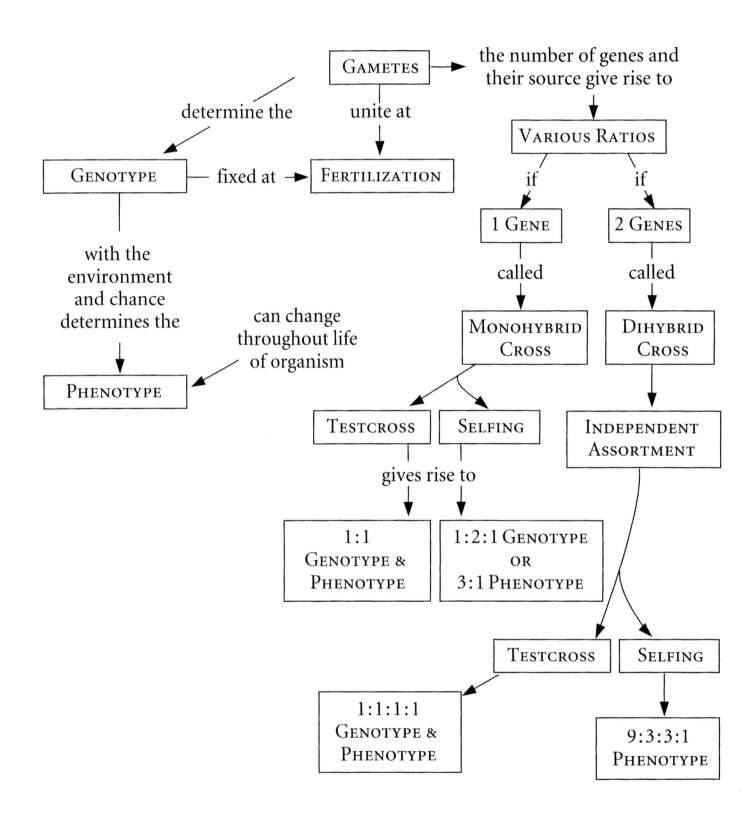

Concept Map 2-1, page 9 in Lavett: *Student Companion with Complete Solutions for IGA, 6e*
Copyright © 1997 W. H. Freeman and Company

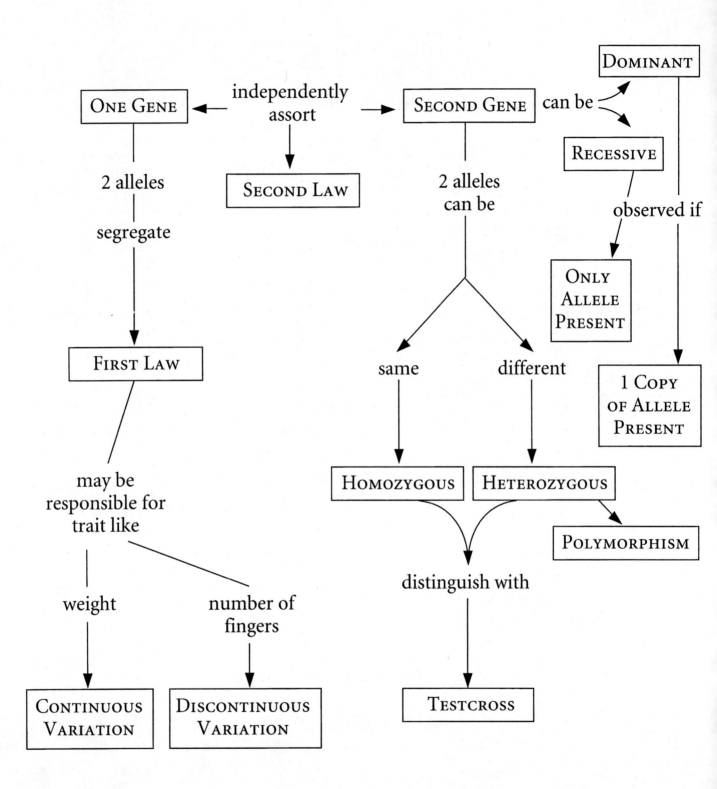

Concept Map 2-2, page 11 in Lavett: *Student Companion with Complete Solutions for IGA*, 6e
Copyright © 1997 W. H. Freeman and Company

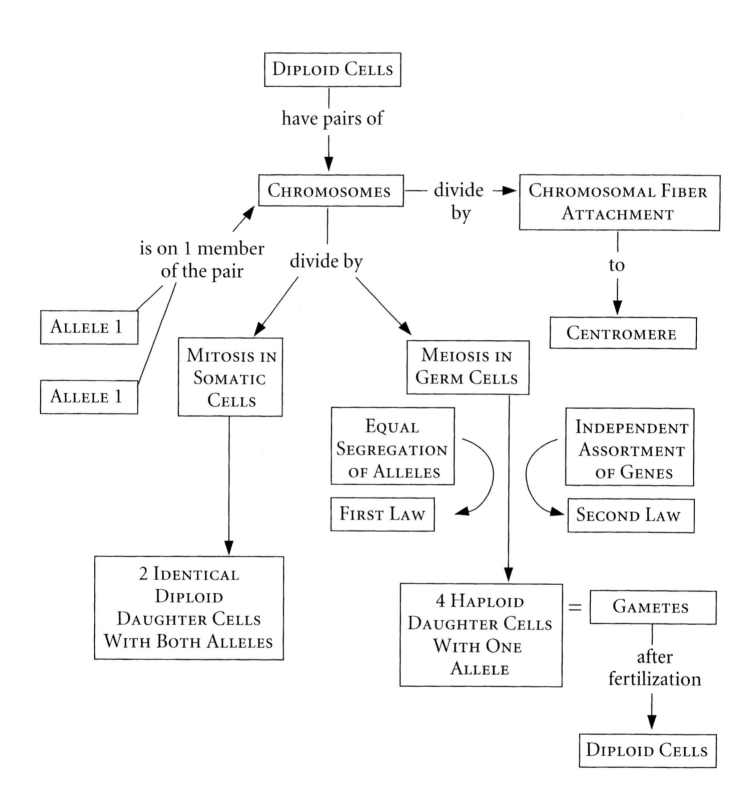

TM-4

Concept Map 3-1, page 33 in Lavett: *Student Companion with Complete Solutions for IGA*, 6e

Copyright © 1997 W. H. Freeman and Company

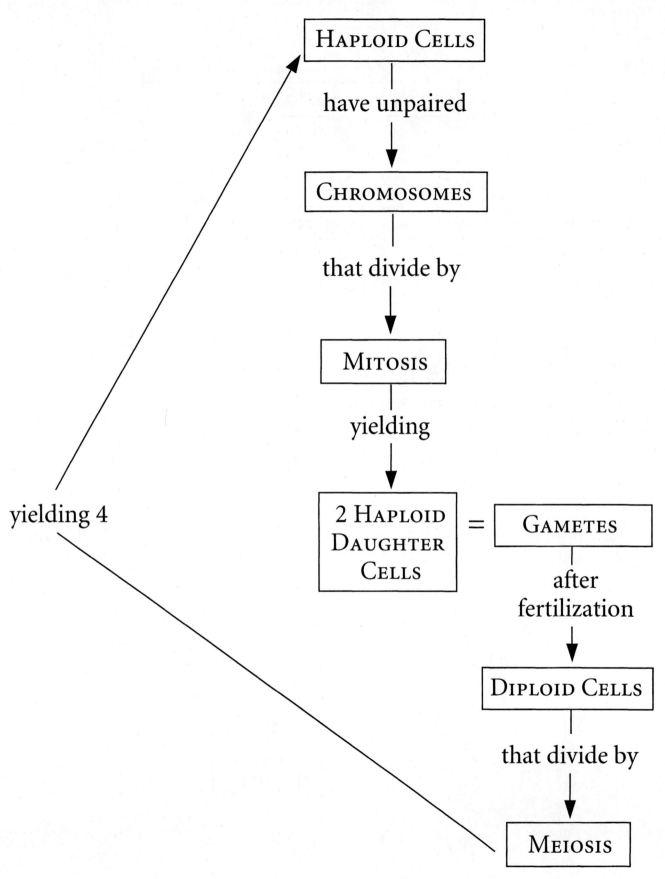

Concept Map 3-2, page 36 in Lavett: *Student Companion with Complete Solutions for IGA*, 6e

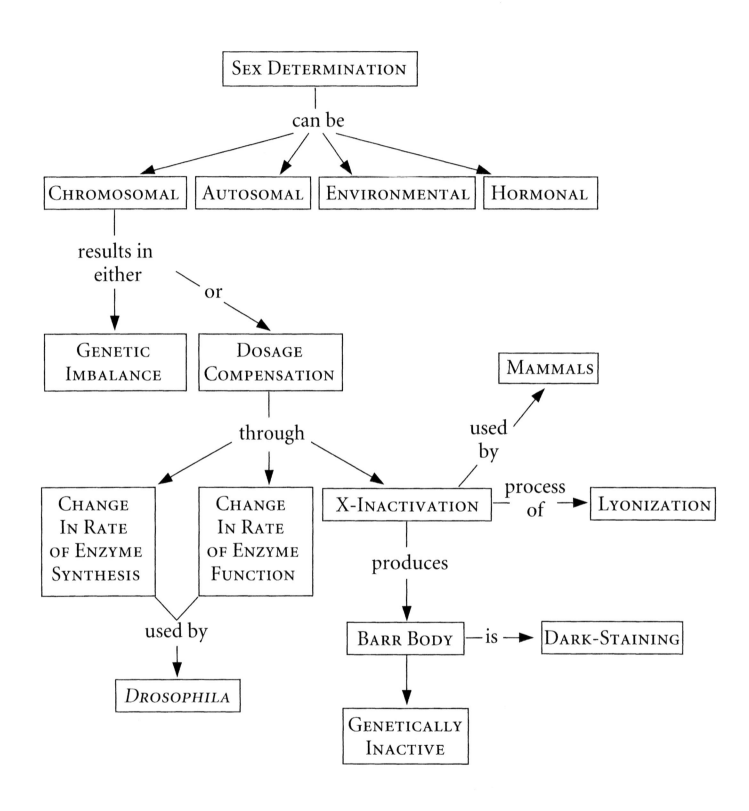

TM-6
Concept Map 3-3, page 39 in Lavett: *Student Companion with Complete Solutions for IGA,* 6e
Copyright © 1997 W. H. Freeman and Company

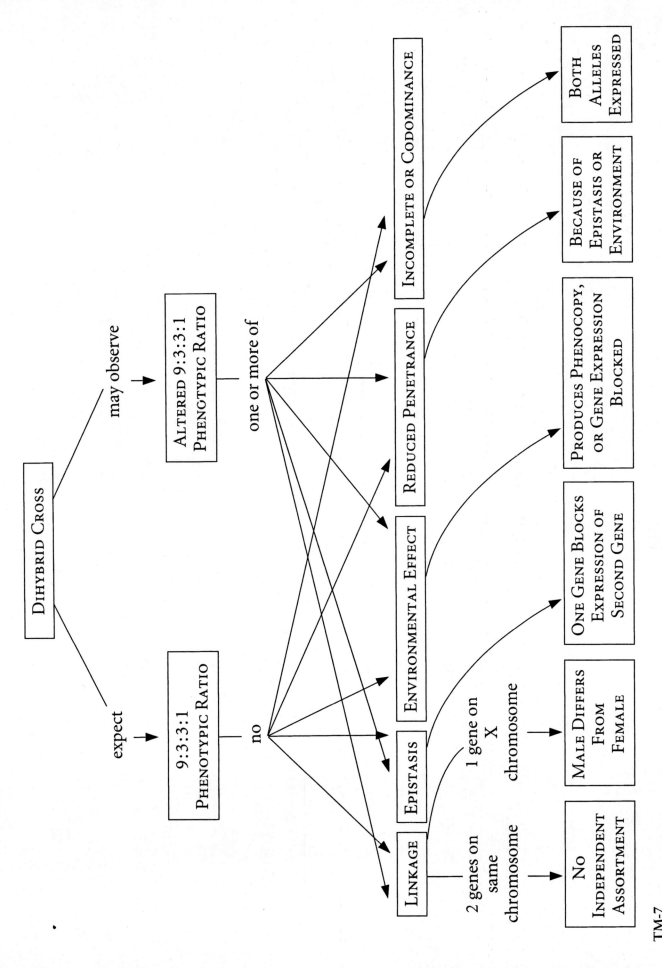

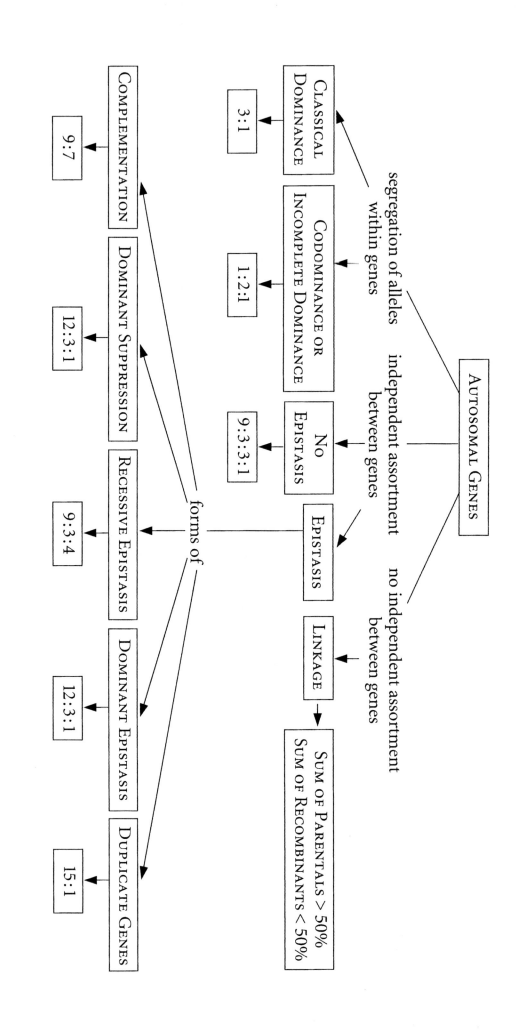

TM-8

Concept Map 4-2, page 56 in Lavett: *Student Companion with Complete Solutions for IGA, 6e*

Copyright © 1997 W. H. Freeman and Company

CLASSICAL DOMINANCE

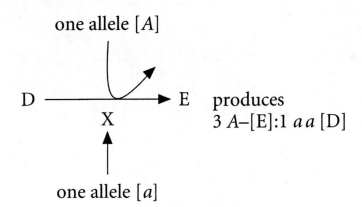

one allele [A]

D ———→ E   produces
    X   3 A–[E]:1 aa [D]

one allele [a]

CODOMINANCE

one allele [A]

D   E   produces   E and F detected
    X   F   1 AA [E]:2 Aa [E + F]:1 aa [D] separately
                                                [ABO blood types]

one allele [a]

INCOMPLETE DOMINANCE

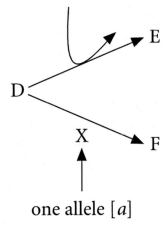

one allele [A]

D   E   produces   E and F detected as
    X   F   1 AA [E]:2 Aa [EF]:1 aa [F]   a mixture
                                                [pink flowers]

one allele [a]

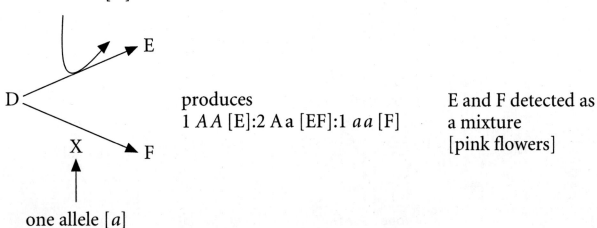

Concept Map 4-3, page 58 in Lavett: *Student Companion with Complete Solutions for IGA, 6e*

NO EPISTASIS

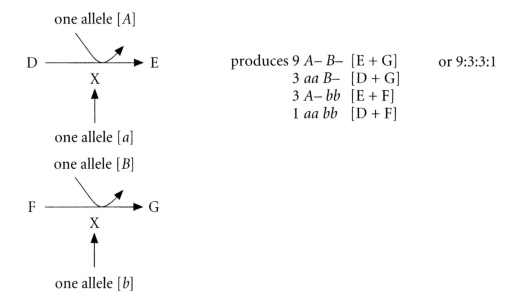

produces 9 *A– B–* [E + G]   or 9:3:3:1
3 *aa B–* [D + G]
3 *A– bb* [E + F]
1 *aa bb* [D + F]

COMPLEMENTATION

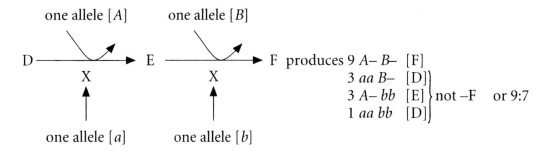

produces 9 *A– B–* [F]
3 *aa B–* [D] ⎫
3 *A– bb* [E] ⎬ not –F   or 9:7
1 *aa bb* [D] ⎭

DOMINANT SUPPRESSION

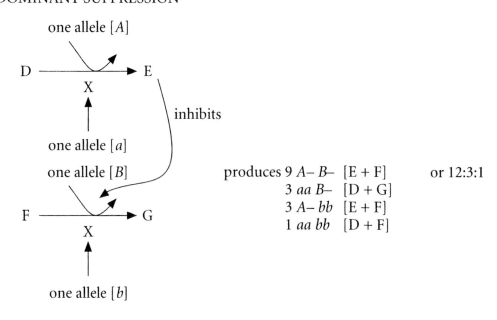

produces 9 *A– B–* [E + F]   or 12:3:1
3 *aa B–* [D + G]
3 *A– bb* [E + F]
1 *aa bb* [D + F]

Concept Map 4-4 (part 1), page 59 in Lavett: *Student Companion with Complete Solutions for IGA,* 6e

RECESSIVE EPISTASIS

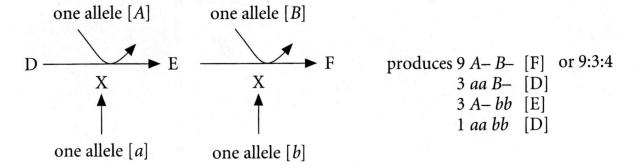

one allele [A]   one allele [B]

D ———→ E ——————→ F   produces 9 *A– B–* [F]   or 9:3:4
         X   X   3 *aa B–* [D]
                                                    3 *A– bb* [E]
                                                    1 *aa bb* [D]

one allele [a]   one allele [b]

DOMINANT EPISTASIS

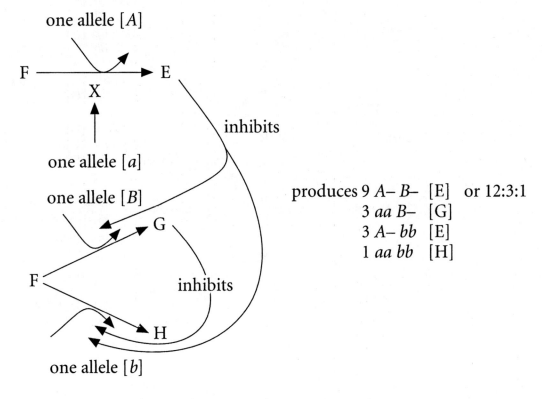

one allele [A]

F ———→ E
        X

one allele [a]

one allele [B]

          inhibits

          G

F

          inhibits

          H

one allele [b]

produces 9 *A– B–* [E]   or 12:3:1
         3 *aa B–* [G]
         3 *A– bb* [E]
         1 *aa bb* [H]

DUPLICATE GENES

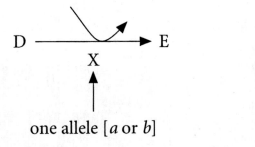

one allele [A or B]

D ——————→ E   produces 9 *A– B–* [E]   or 15:1
        X   3 *aa B–* [E]
                                3 *A– bb* [E]
                                1 *aa bb* [D]

one allele [a or b]

Concept Map 4-4 (part 2), page 60 in Lavett: *Student Companion with Complete Solutions for IGA,* 6e

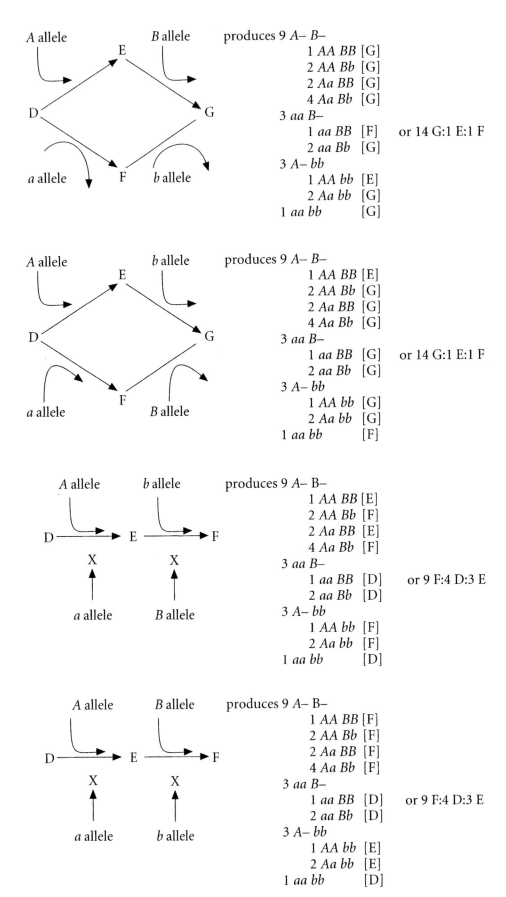

produces 9 A– B–
1 *AA BB* [G]
2 *AA Bb* [G]
2 *Aa BB* [G]
4 *Aa Bb* [G]
3 *aa B–*
1 *aa BB* [F]   or 14 G:1 E:1 F
2 *aa Bb* [G]
3 *A– bb*
1 *AA bb* [E]
2 *Aa bb* [G]
1 *aa bb* [G]

produces 9 A– B–
1 *AA BB* [E]
2 *AA Bb* [G]
2 *Aa BB* [G]
4 *Aa Bb* [G]
3 *aa B–*
1 *aa BB* [G]   or 14 G:1 E:1 F
2 *aa Bb* [G]
3 *A– bb*
1 *AA bb* [G]
2 *Aa bb* [G]
1 *aa bb* [F]

produces 9 A– B–
1 *AA BB* [E]
2 *AA Bb* [F]
2 *Aa BB* [E]
4 *Aa Bb* [F]
3 *aa B–*
1 *aa BB* [D]   or 9 F:4 D:3 E
2 *aa Bb* [D]
3 *A– bb*
1 *AA bb* [F]
2 *Aa bb* [F]
1 *aa bb* [D]

produces 9 A– B–
1 *AA BB* [F]
2 *AA Bb* [F]
2 *Aa BB* [F]
4 *Aa Bb* [F]
3 *aa B–*
1 *aa BB* [D]   or 9 F:4 D:3 E
2 *aa Bb* [D]
3 *A– bb*
1 *AA bb* [E]
2 *Aa bb* [E]
1 *aa bb* [D]

Concept Map 4-5, page 62 in Lavett: *Student Companion with Complete Solutions for IGA*, 6e

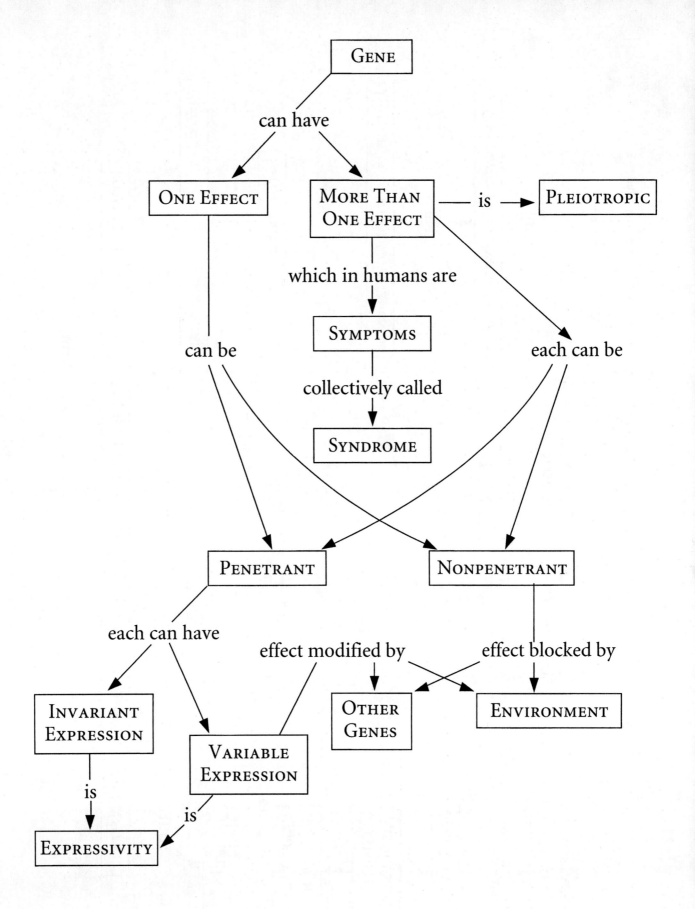

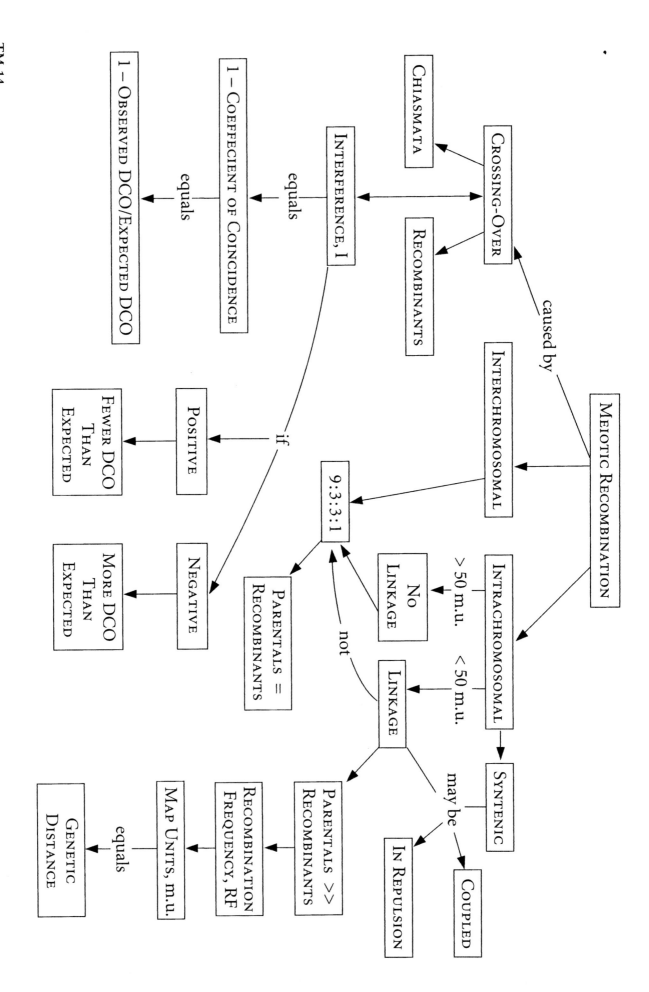

TM-14
Concept Map 5-1, page 101 in Lavett: *Student Companion with Complete Solutions for IGA, 6e*
Copyright © 1997 W. H. Freeman and Company

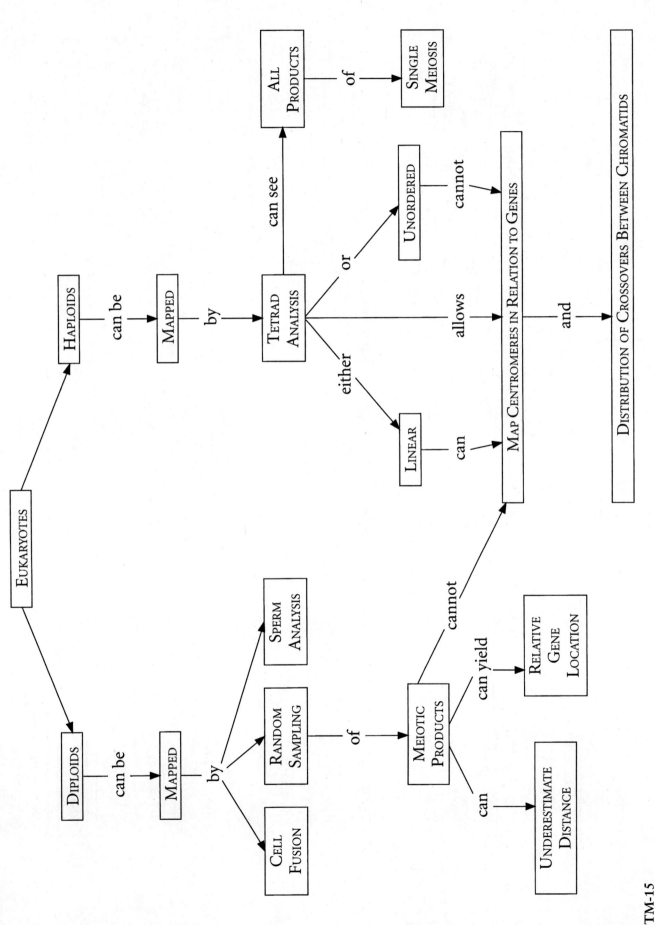

TM-15

Concept Map 6-1, page 130 in Lavett: *Student Companion with Complete*
Solutions for IGA, 6e
Copyright © 1997 W. H. Freeman and Company

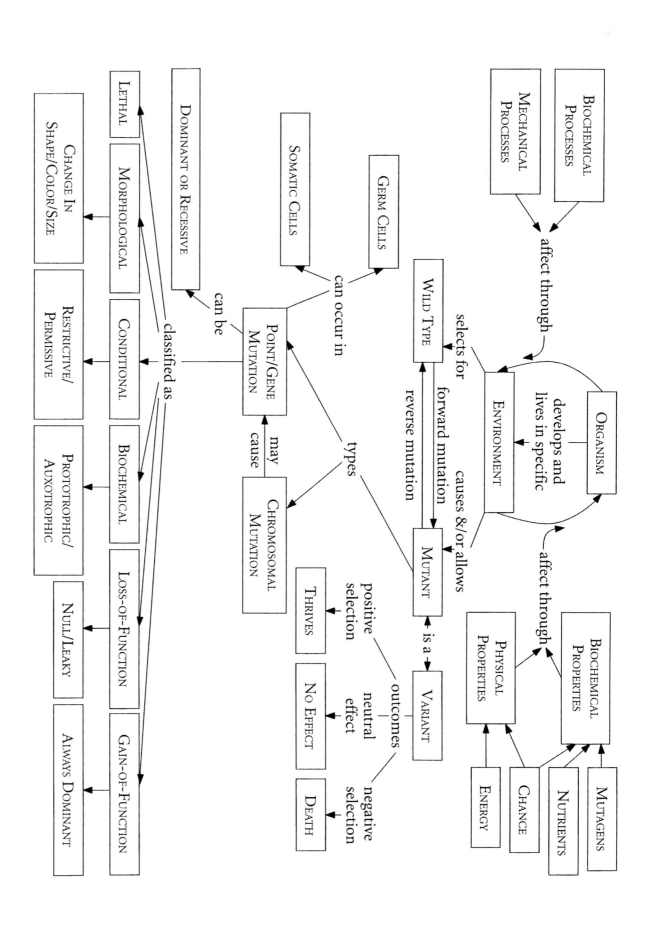

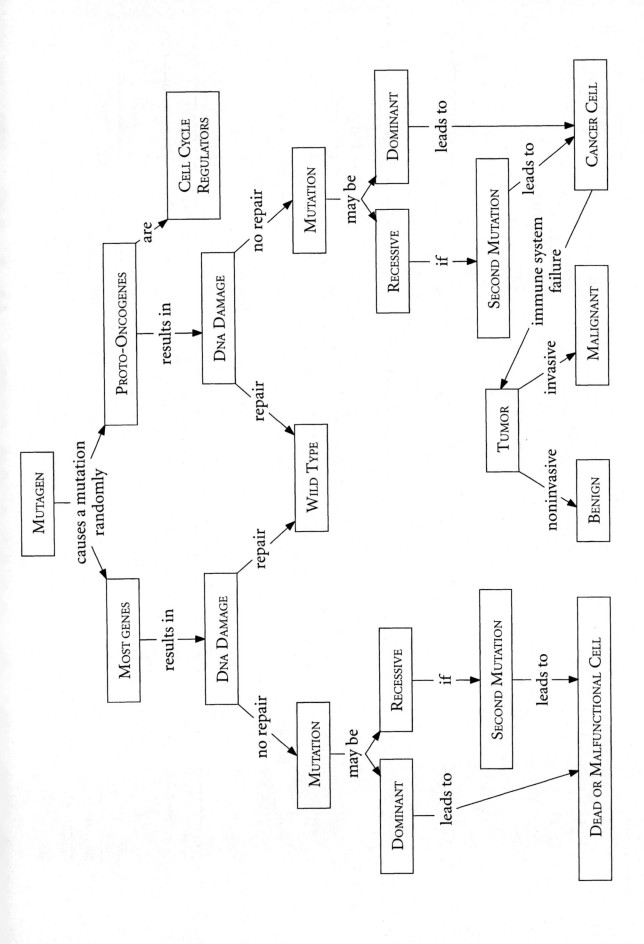

TM-17

Concept Map 7-2, page 166 in Lavett: *Student Companion with Complete Solutions for IGA, 6e*

Copyright © 1997 W. H. Freeman and Company

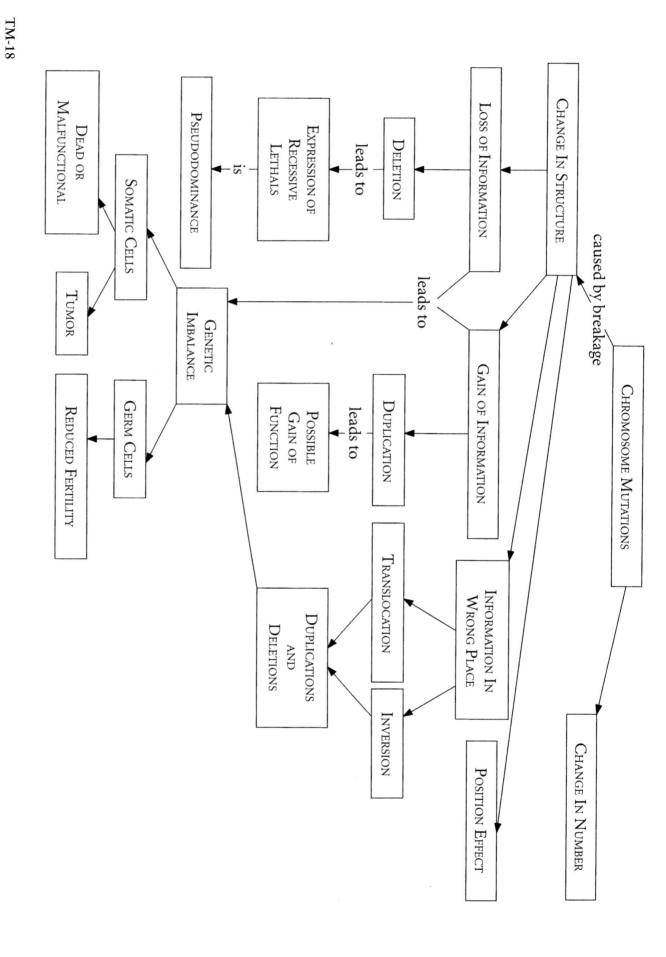

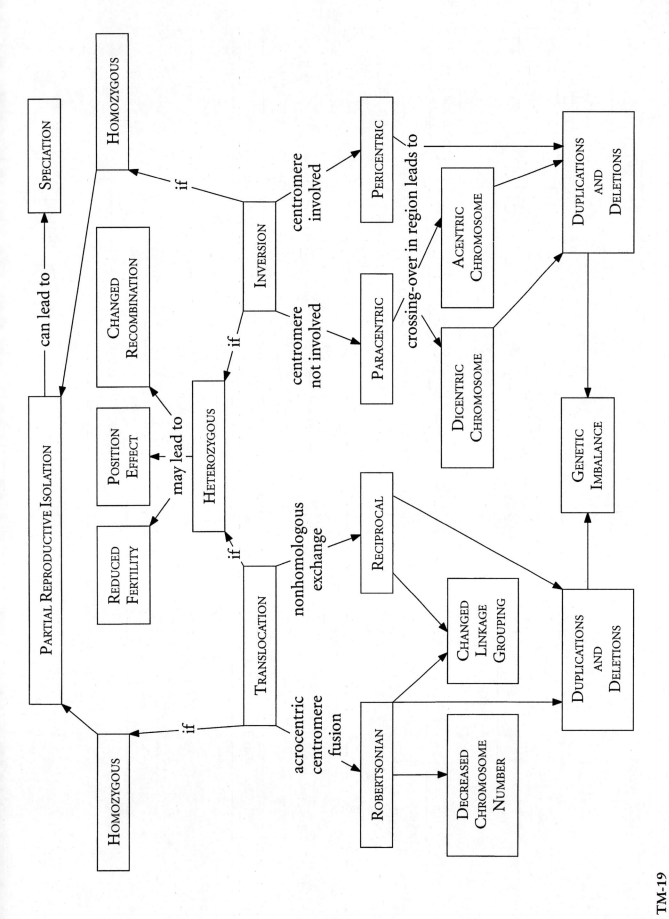

TM-19

Concept Map 8-2, page 182 in Lavett: *Student Companion with Complete Solutions for IGA*, 6e

Copyright © 1997 W. H. Freeman and Company

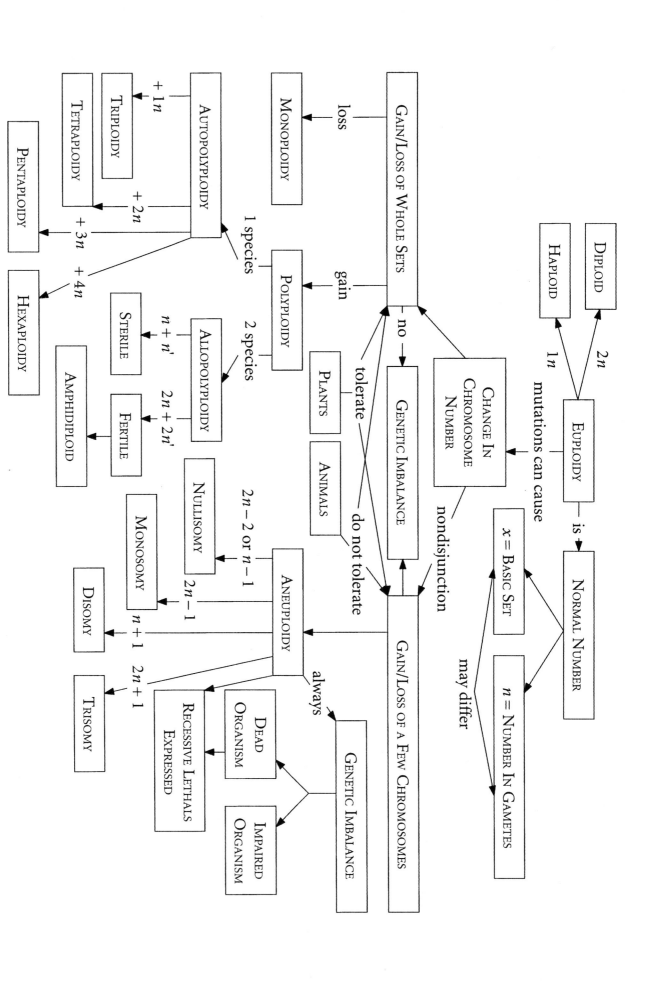

TM-20
Concept Map 9-1, page 208 in Lavett: *Student Companion with Complete Solutions for IGA, 6e*
Copyright © 1997 W. H. Freeman and Company

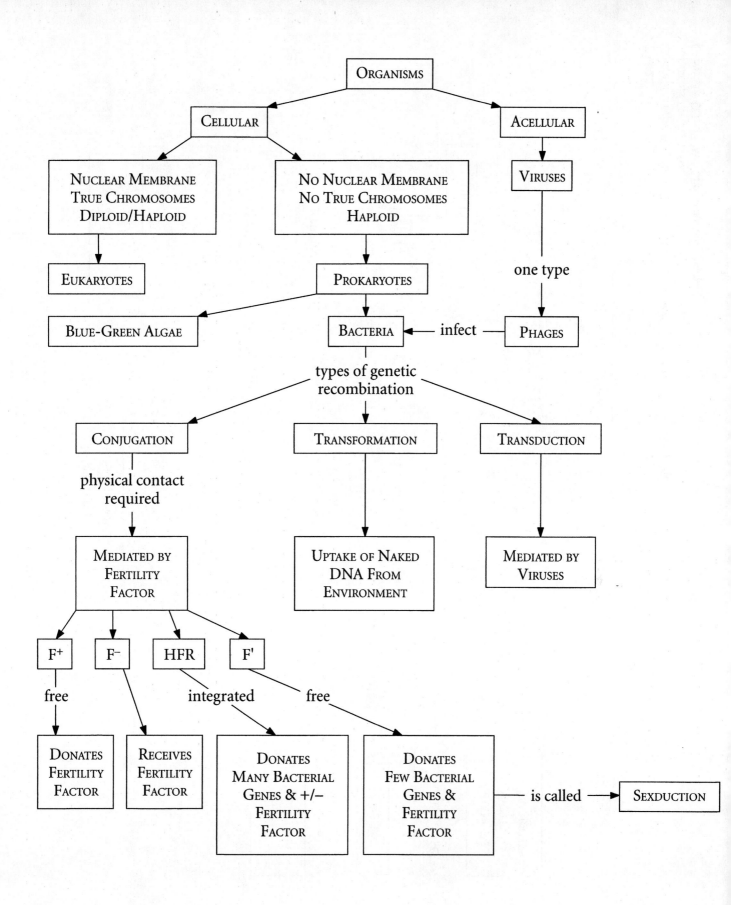

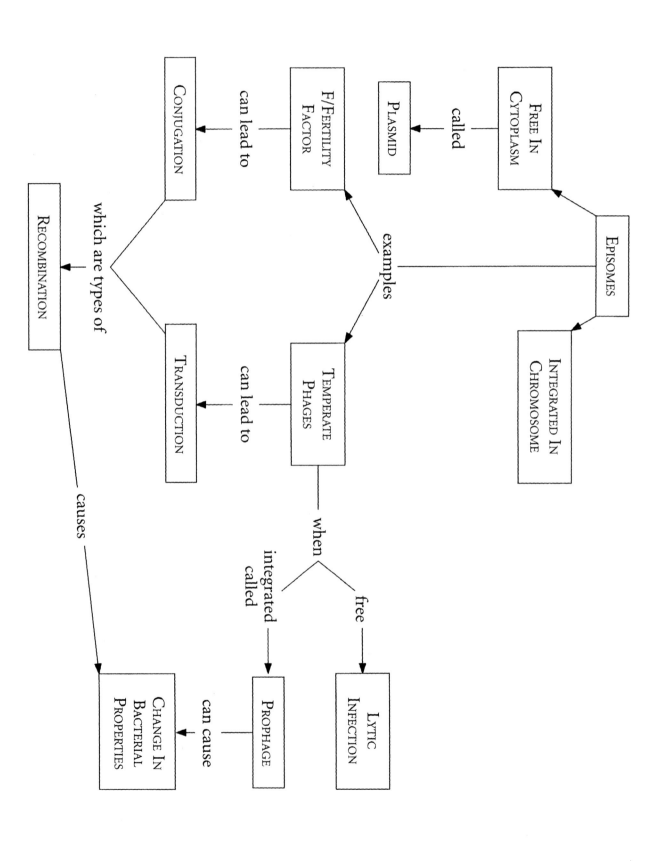

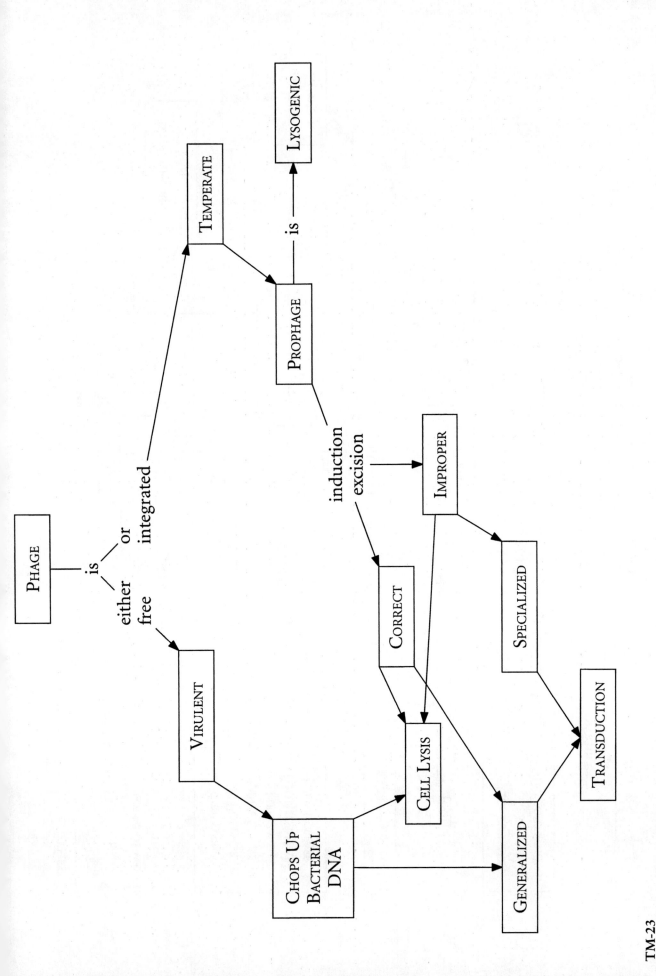

Concept Map 10-3, page 230 in Lavett: *Student Companion with Complete Solutions for IGA*, 6e
Copyright © 1997 W. H. Freeman and Company

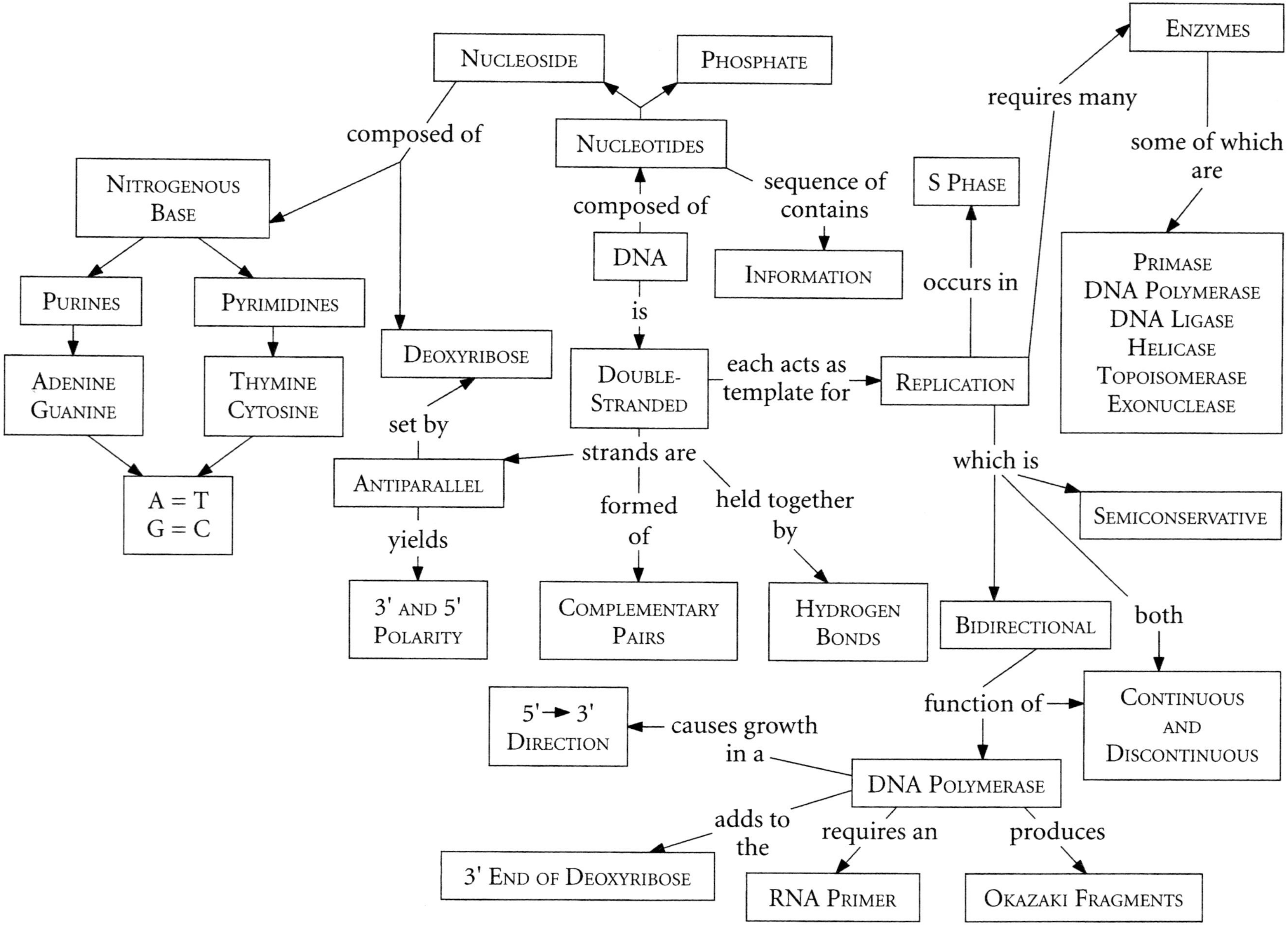

TM-24
Concept Map 11-1, page 242 in Lavett: *Student Companion with Complete Solutions for IGA*, 6e

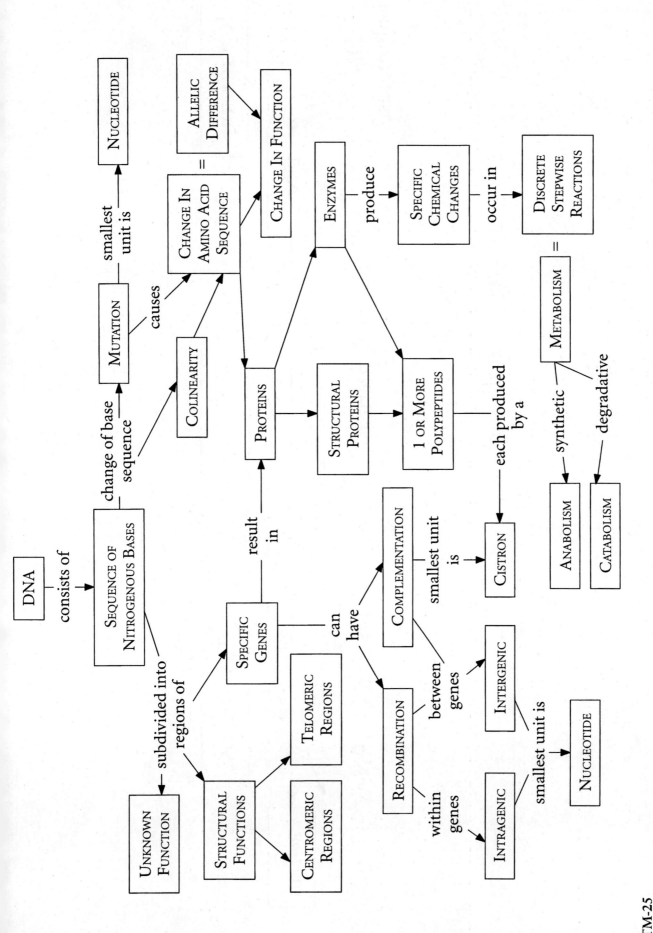

Concept Map 12-1, page 254 in Lavett: *Student Companion with Complete Solutions for IGA, 6e*
Copyright © 1997 W. H. Freeman and Company

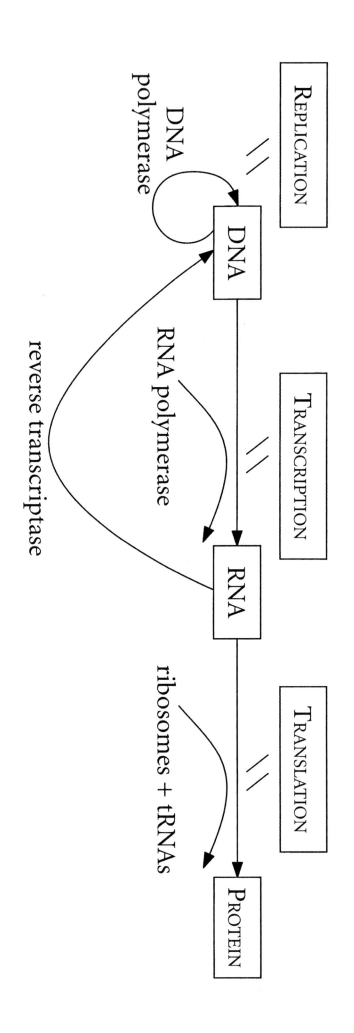

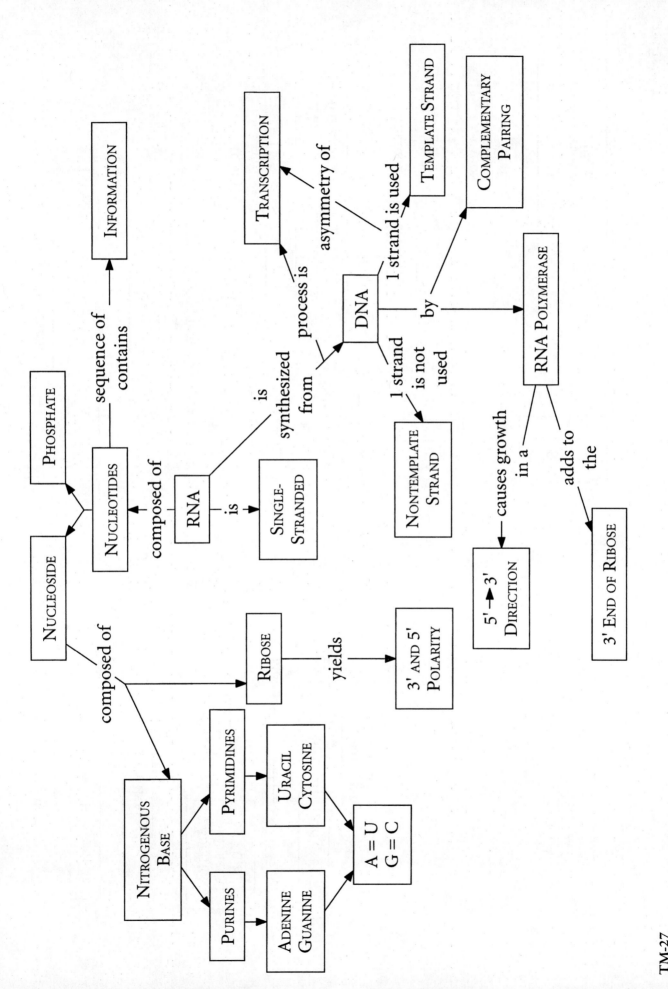

Concept Map 13-2, page 272 in Lavett: *Student Companion with Complete Solutions for IGA, 6e*
Copyright © 1997 W. H. Freeman and Company

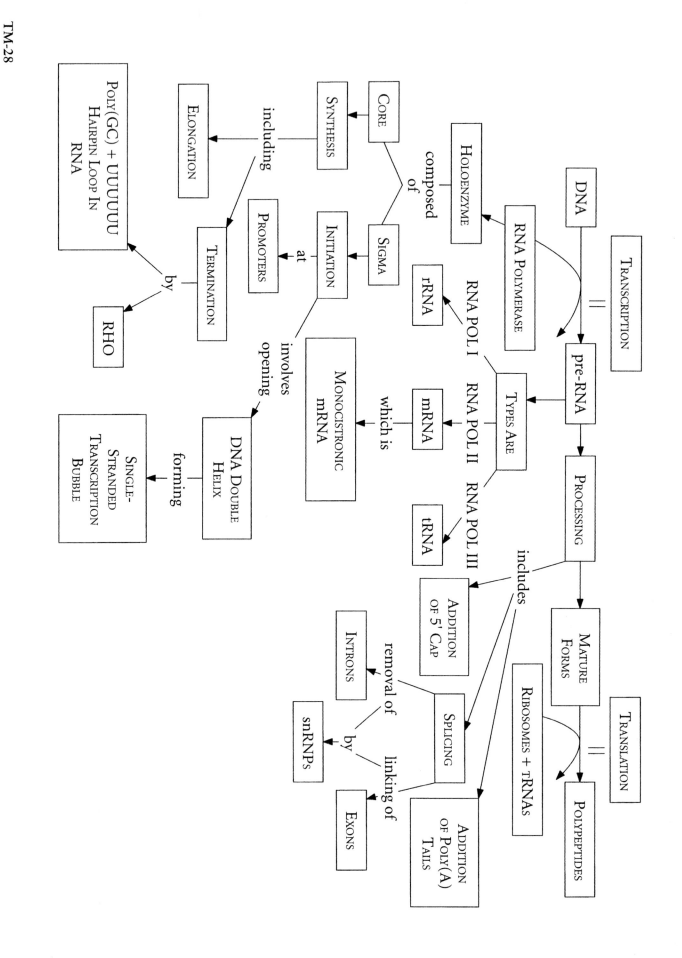

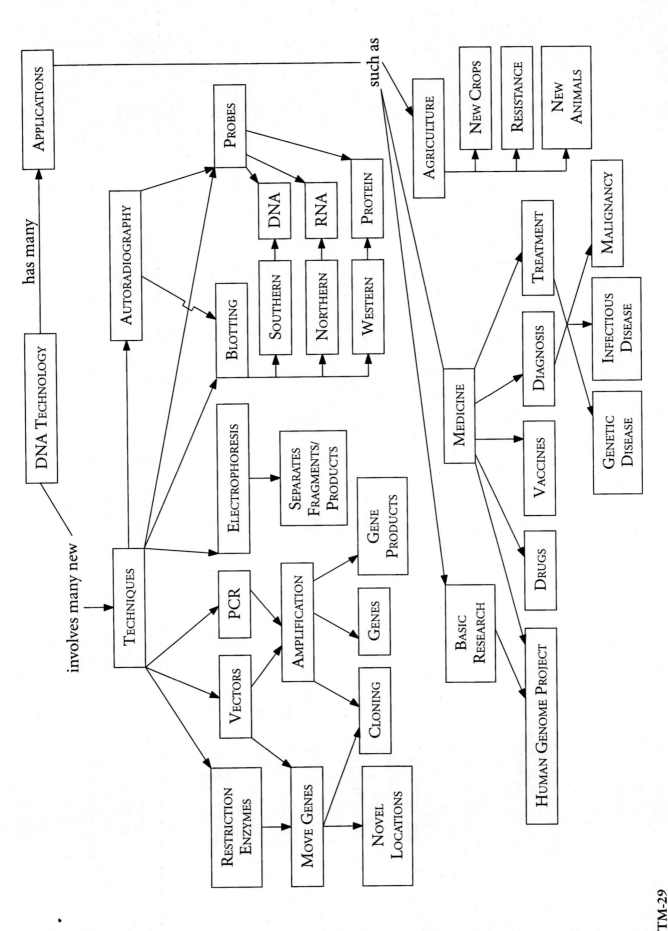

TM-29
Concept Map 14-1, page 282 in Lavett: *Student Companion with Complete Solutions for IGA, 6e*
Copyright © 1997 W. H. Freeman and Company

TM-30

Concept Map 15-1, page 296 in Lavett: *Student Companion with Complete Solutions for IGA, 6e*

Copyright © 1997 W. H. Freeman and Company

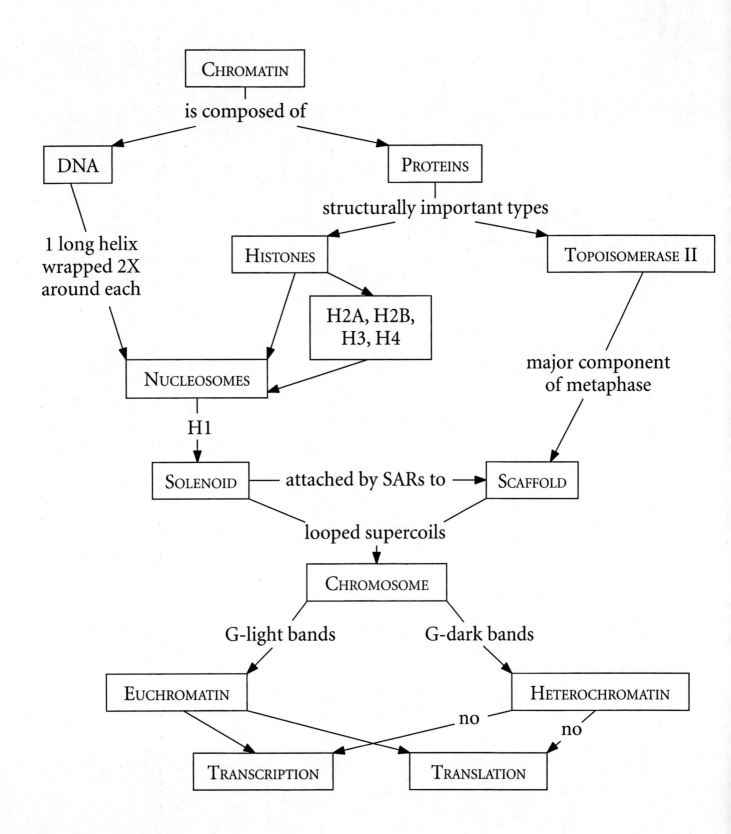

Concept Map 16-1 page 310 in Lavett: *Student Companion with Complete Solutions for IGA,* 6e
Copyright © 1997 W. H. Freeman and Company

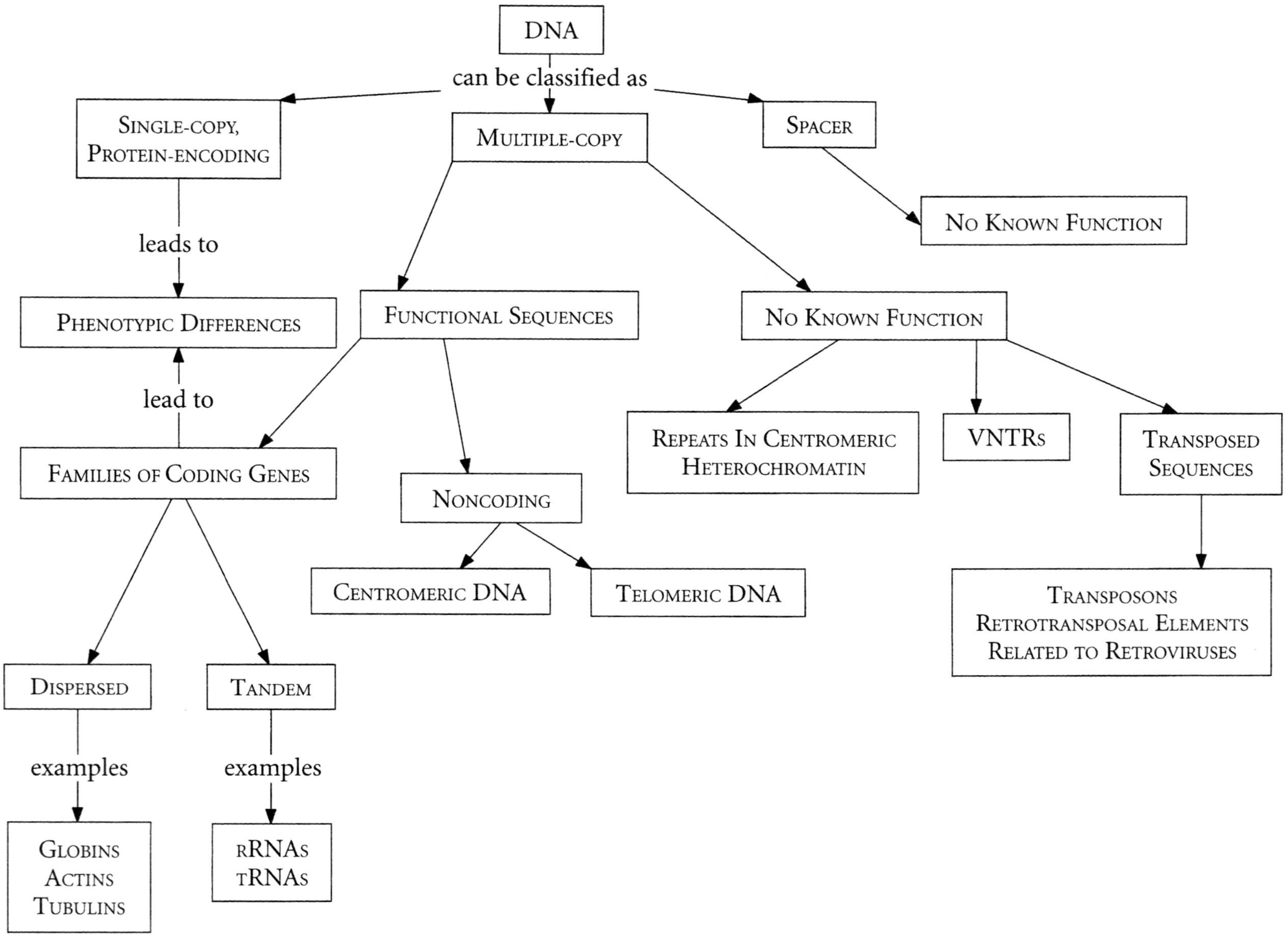

DNA

can be classified as

SINGLE-COPY, PROTEIN-ENCODING

MULTIPLE-COPY

SPACER

leads to

PHENOTYPIC DIFFERENCES

lead to

FAMILIES OF CODING GENES

FUNCTIONAL SEQUENCES

NO KNOWN FUNCTION

NO KNOWN FUNCTION

DISPERSED

examples

GLOBINS
ACTINS
TUBULINS

TANDEM

examples

rRNAs
tRNAs

NONCODING

CENTROMERIC DNA

TELOMERIC DNA

REPEATS IN CENTROMERIC HETEROCHROMATIN

VNTRs

TRANSPOSED SEQUENCES

TRANSPOSONS
RETROTRANSPOSAL ELEMENTS
RELATED TO RETROVIRUSES

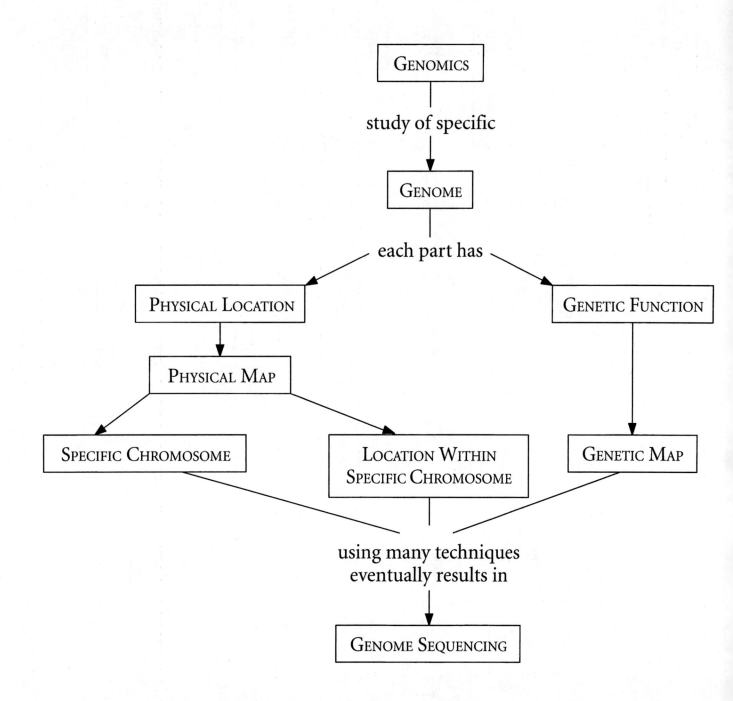

Concept Map 17-1, page 324 in Lavett: *Student Companion with Complete Solutions for IGA,* 6e

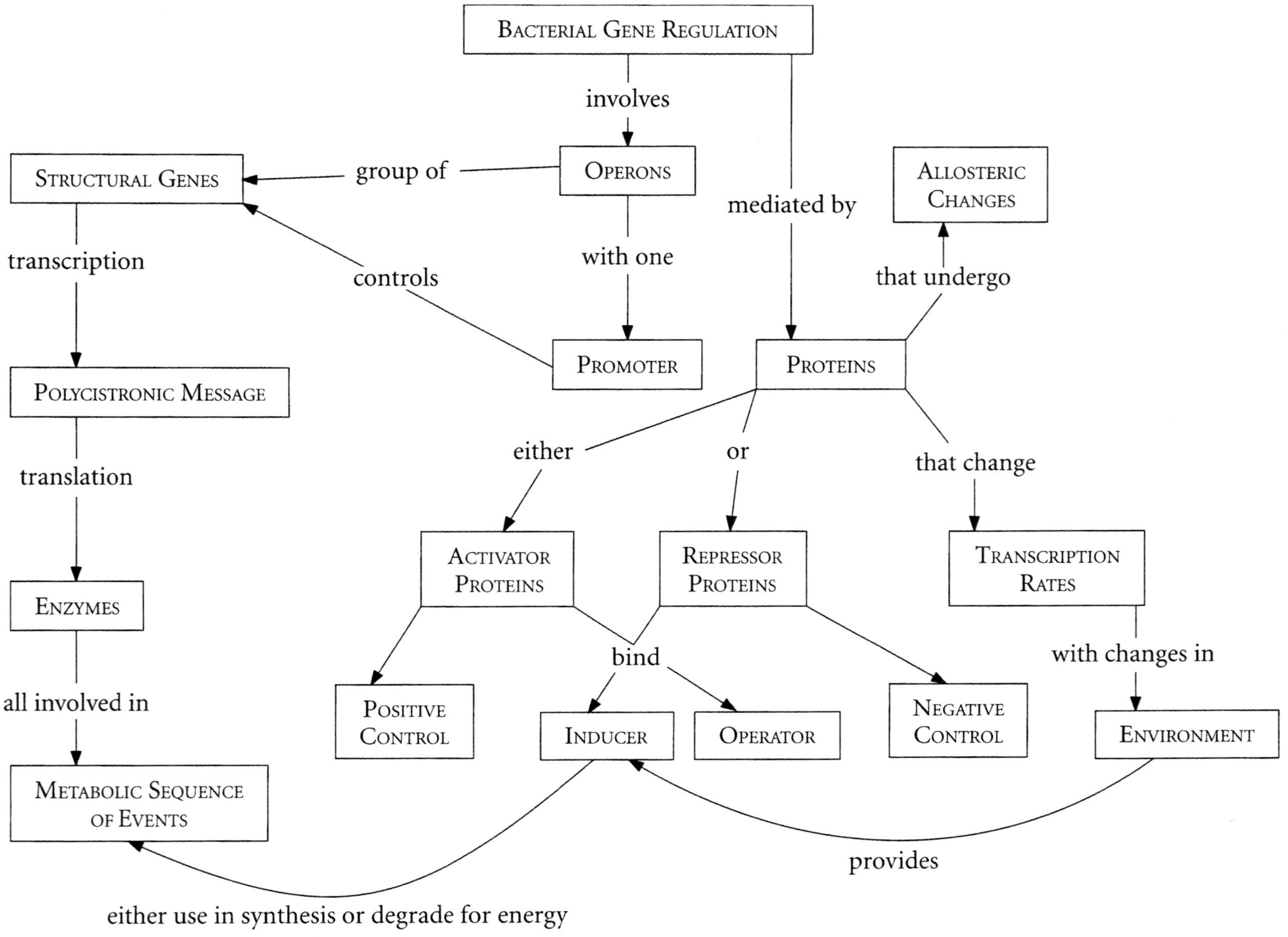

TM-34

Concept Map 18-1, page 336 in Lavett: *Student Companion with Complete Solutions for IGA, 6e*

Copyright © 1997 W. H. Freeman and Company

STRUCTURAL GENES

transcription

POLYCISTRONIC MESSAGE

translation

ENZYMES

all involved in

METABOLIC SEQUENCE OF EVENTS

either use in synthesis or degrade for energy

controls

group of

BACTERIAL GENE REGULATION

involves

OPERONS

with one

PROMOTER

either

ACTIVATOR PROTEINS

POSITIVE CONTROL

bind

INDUCER

OPERATOR

REPRESSOR PROTEINS

NEGATIVE CONTROL

or

PROTEINS

mediated by

that undergo

ALLOSTERIC CHANGES

that change

TRANSCRIPTION RATES

with changes in

ENVIRONMENT

provides

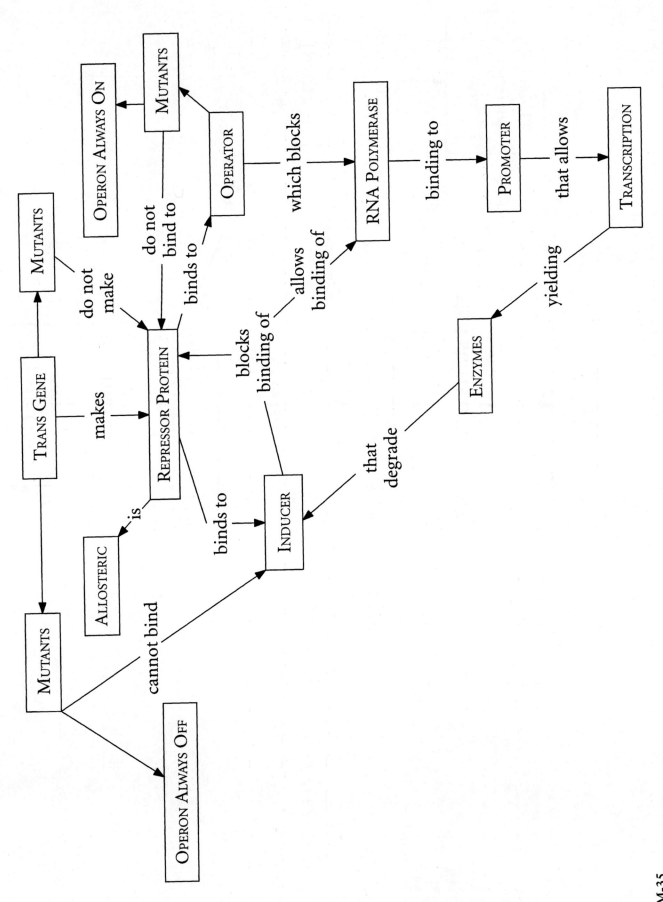

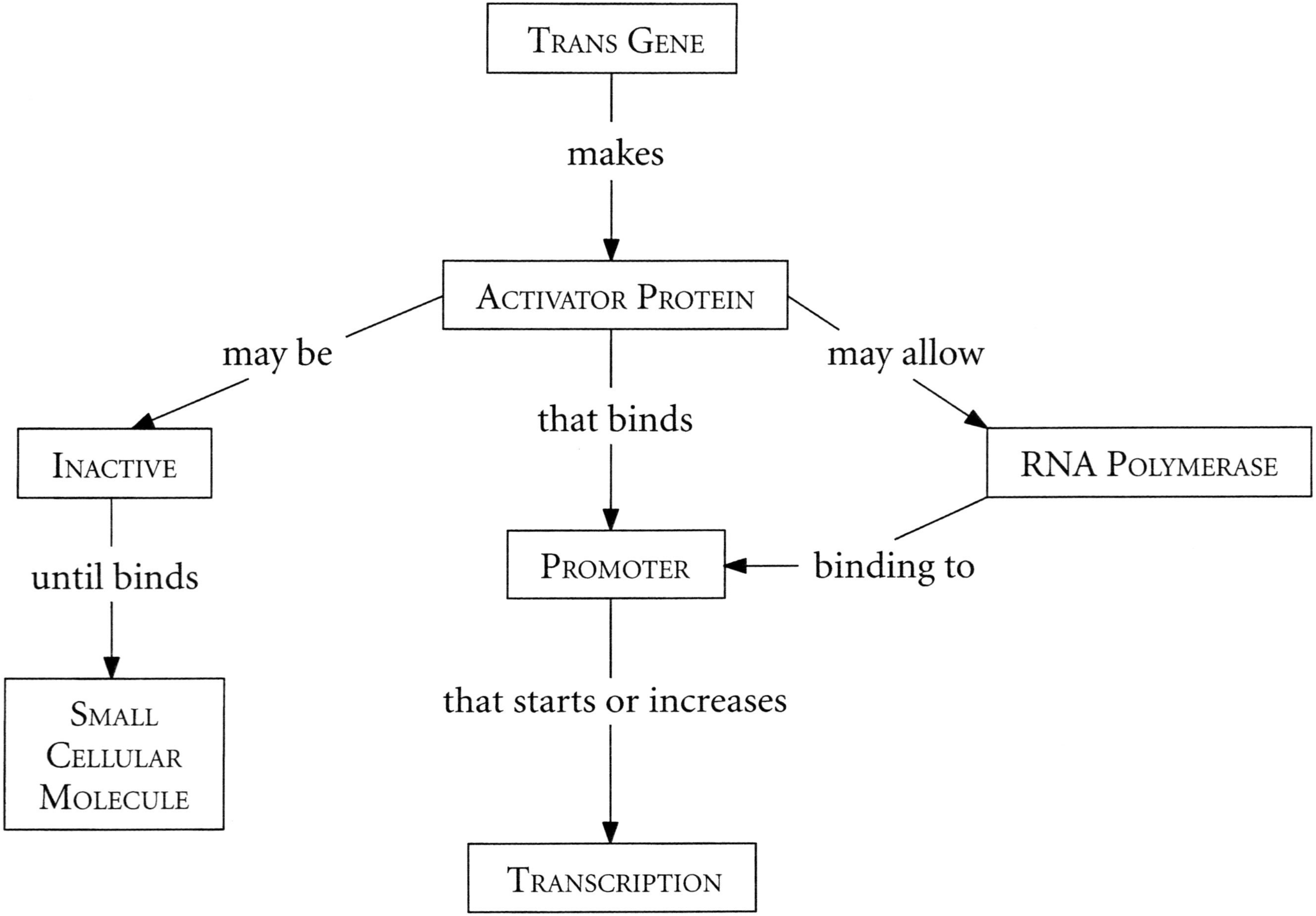

TM-36
Concept Map 18-3, page 340 in Lavett: *Student Companion with Complete Solutions for IGA, 6e*
Copyright © 1997 W. H. Freeman and Company

SMALL CELLULAR MOLECULE

until binds

INACTIVE

may be

ACTIVATOR PROTEIN

makes

TRANS GENE

that binds

PROMOTER

that starts or increases

TRANSCRIPTION

binding to

may allow

RNA POLYMERASE

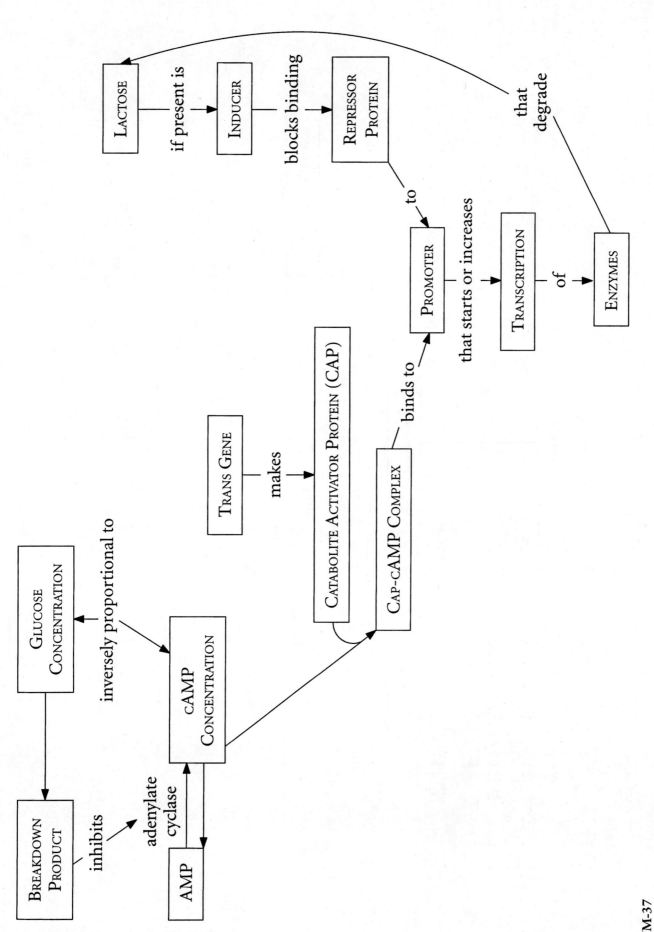

TM-37

Concept Map 18-4, page 341 in Lavett: *Student Companion with Complete Solutions for IGA, 6e*

Copyright © 1997 W. H. Freeman and Company

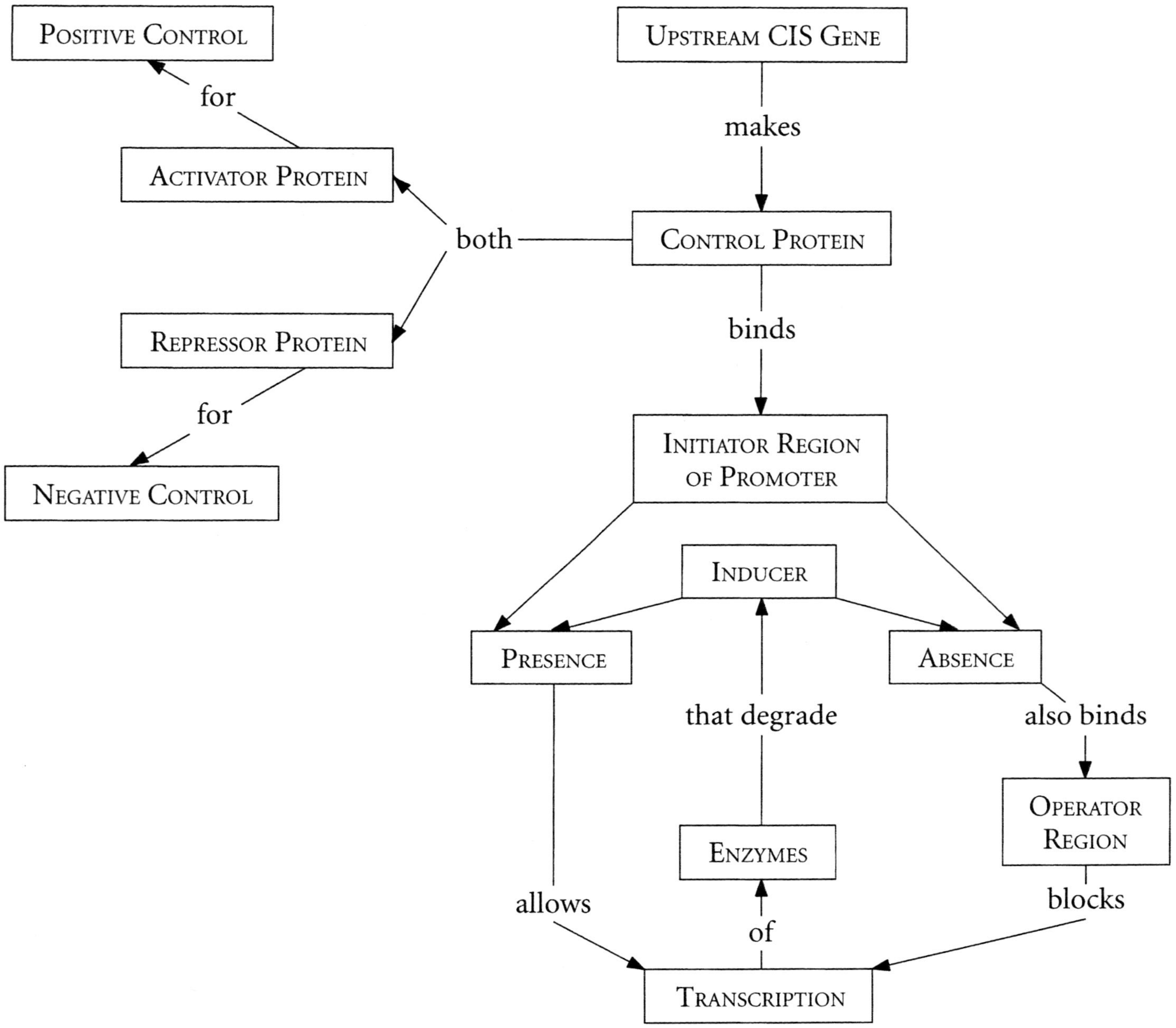

NEGATIVE CONTROL

REPRESSOR PROTEIN

for

ACTIVATOR PROTEIN

for

POSITIVE CONTROL

both

PRESENCE

allows

INDUCER

INITIATOR REGION OF PROMOTER

binds

CONTROL PROTEIN

makes

UPSTREAM CIS GENE

ENZYMES

of

that degrade

TRANSCRIPTION

ABSENCE

also binds

OPERATOR REGION

blocks

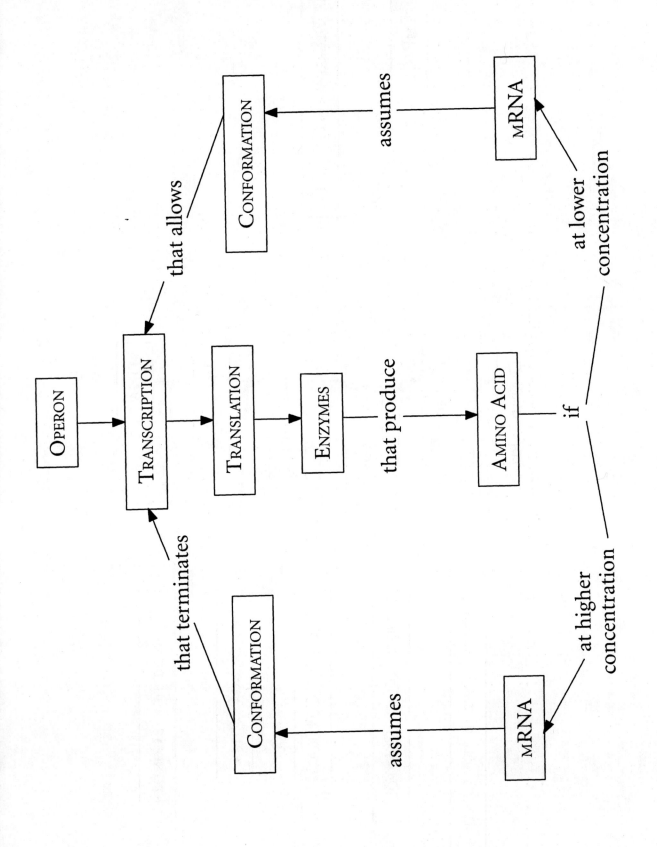

Concept Map 18-6, page 344 in Lavett: *Student Companion with Complete*
Solutions for IGA, 6e

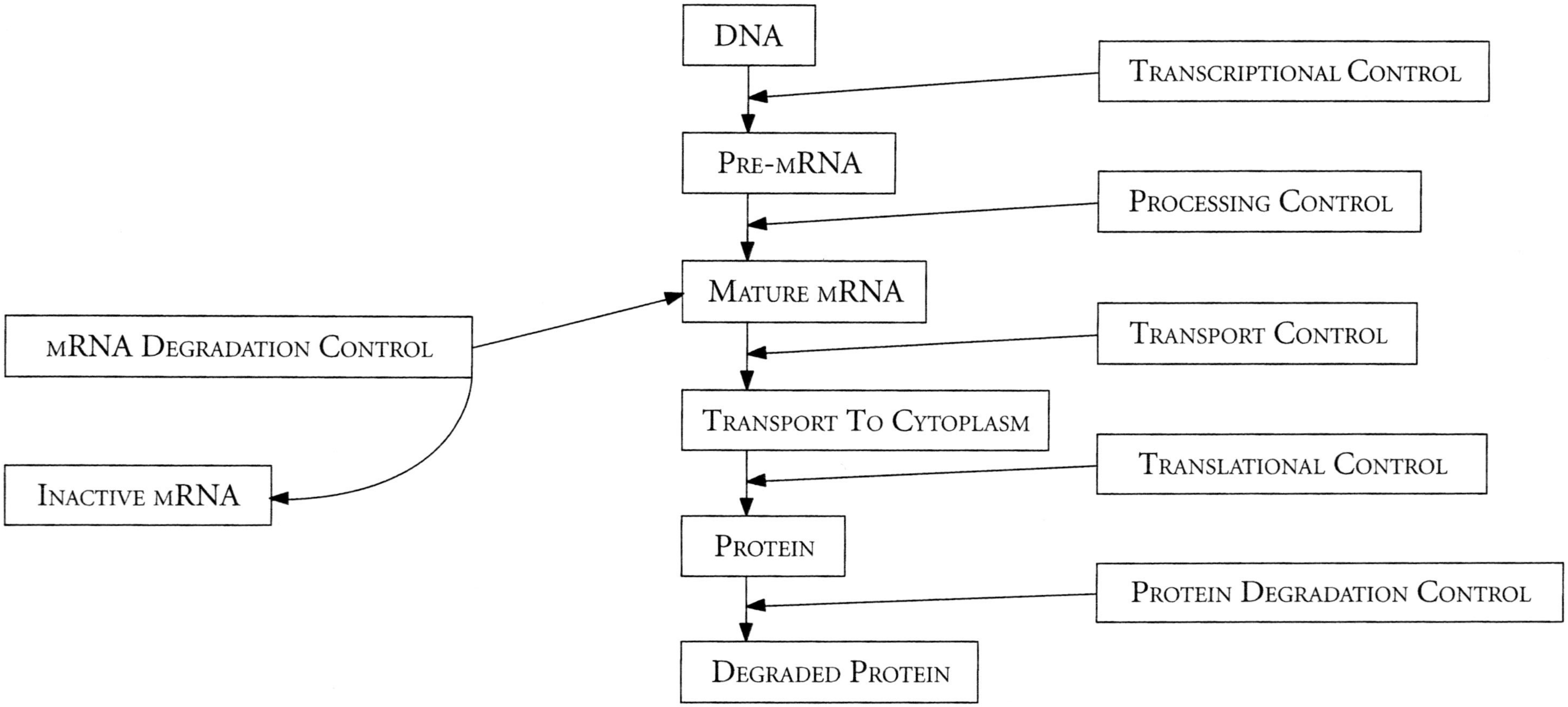

TM-40
Concept Map 18-7, page 345 in Lavett: Student Companion with Complete
Solutions for IGA, 6e
Copyright © 1997 W. H. Freeman and Company

INACTIVE mRNA

mRNA DEGRADATION CONTROL

DNA

PRE-mRNA

MATURE mRNA

TRANSPORT TO CYTOPLASM

PROTEIN

DEGRADED PROTEIN

TRANSCRIPTIONAL CONTROL

PROCESSING CONTROL

TRANSPORT CONTROL

TRANSLATIONAL CONTROL

PROTEIN DEGRADATION CONTROL

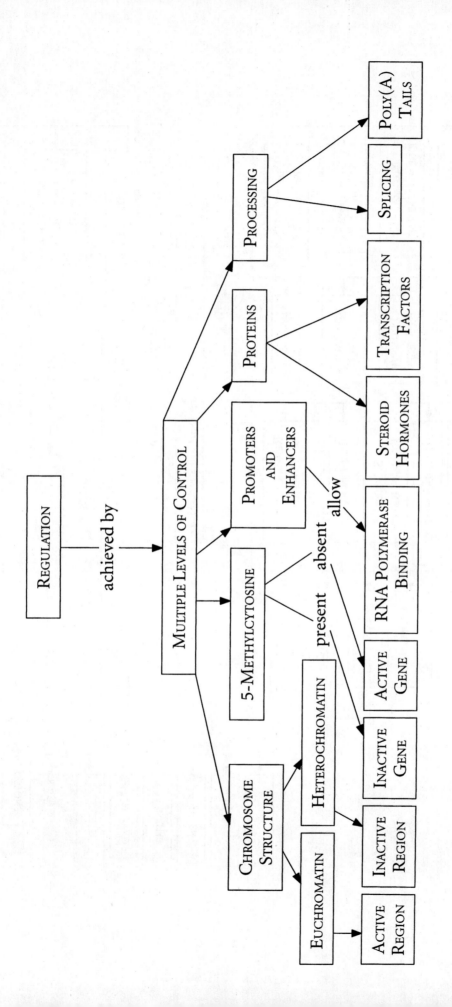

TM-41

Concept Map 18-8, page 347 in Lavett: *Student Companion with Complete Solutions for IGA*, 6e

Copyright © 1997 W. H. Freeman and Company

TM-42
Concept Map 19-1, page 354 in Lavett: *Student Companion with Complete Solutions for IGA, 6e*
Copyright © 1997 W. H. Freeman and Company

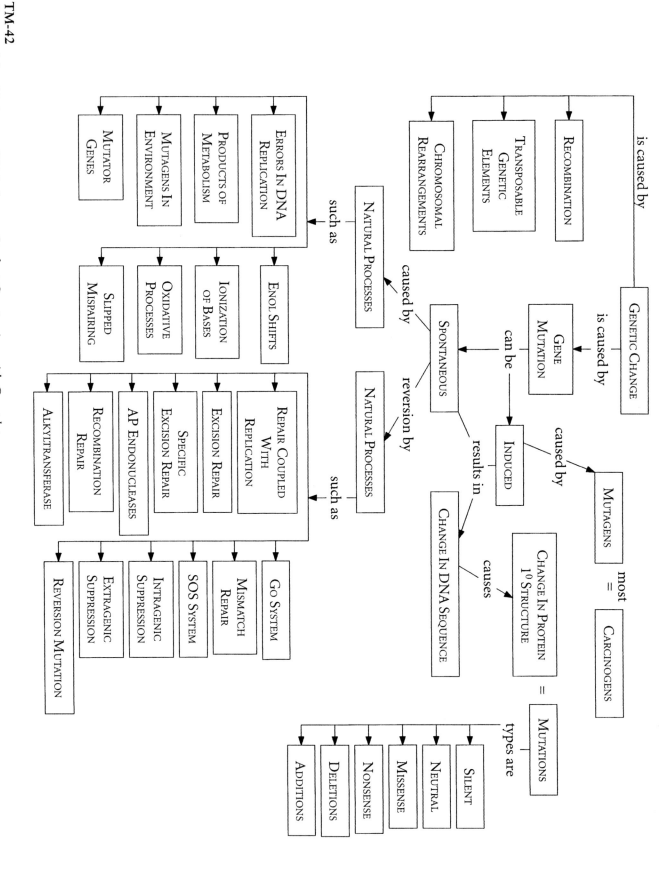

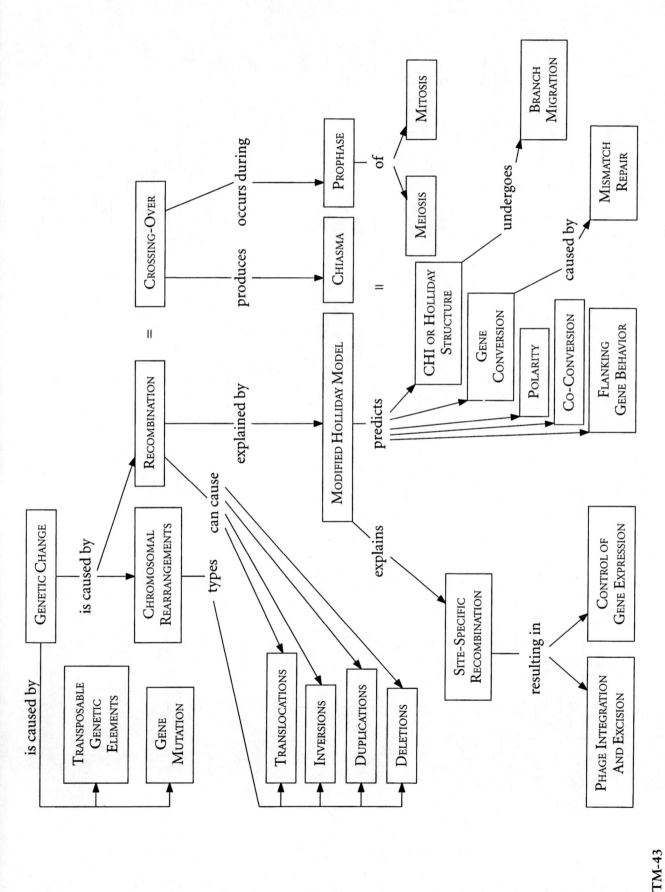

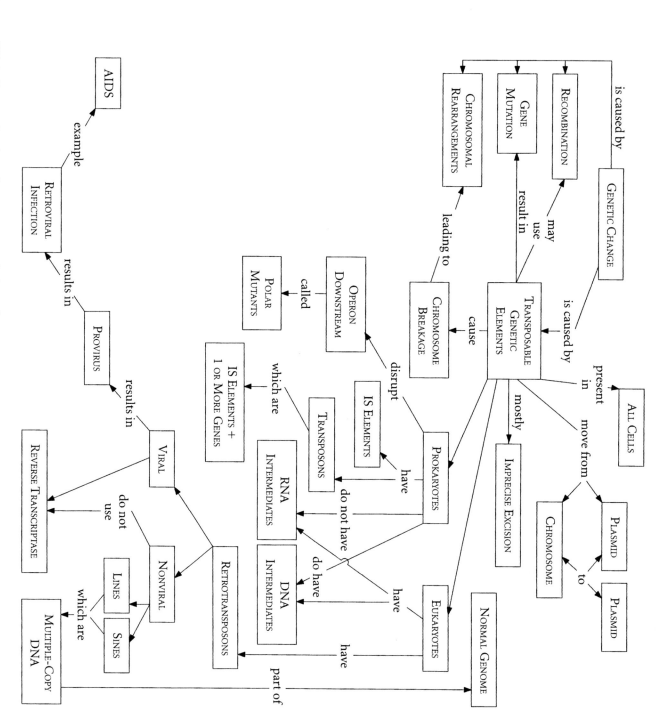

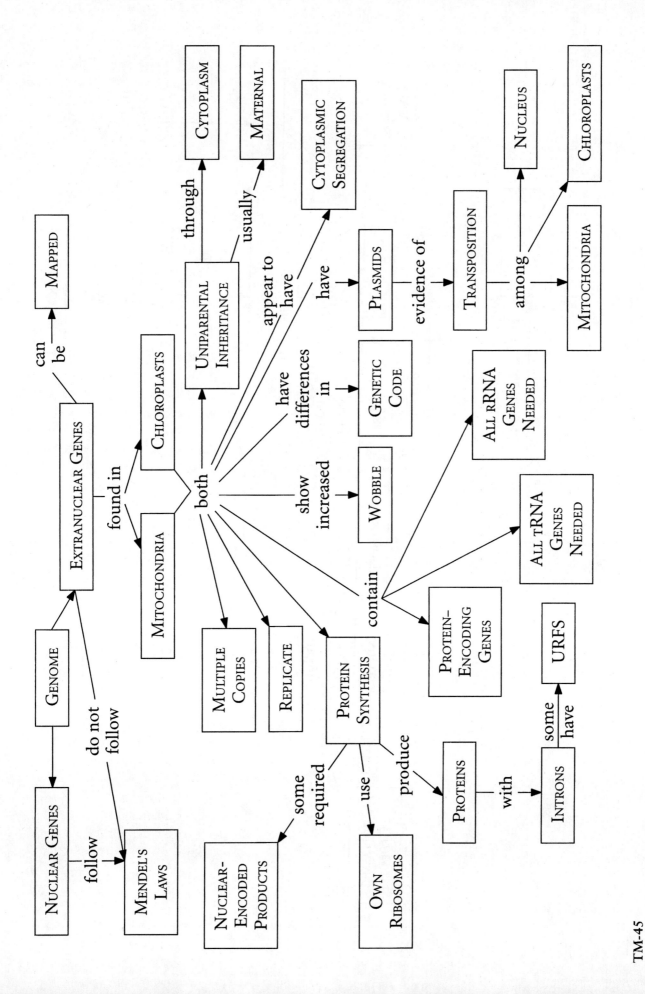

Concept Map 22-1, page 378 in Lavett: *Student Companion with Complete Solutions for IGA, 6e*
Copyright © 1997 W. H. Freeman and Company

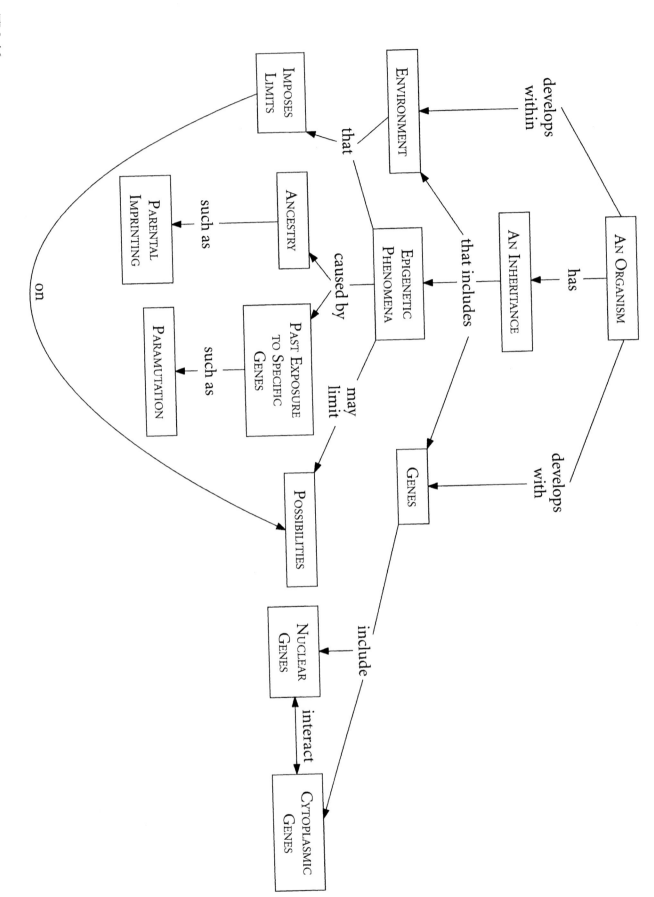

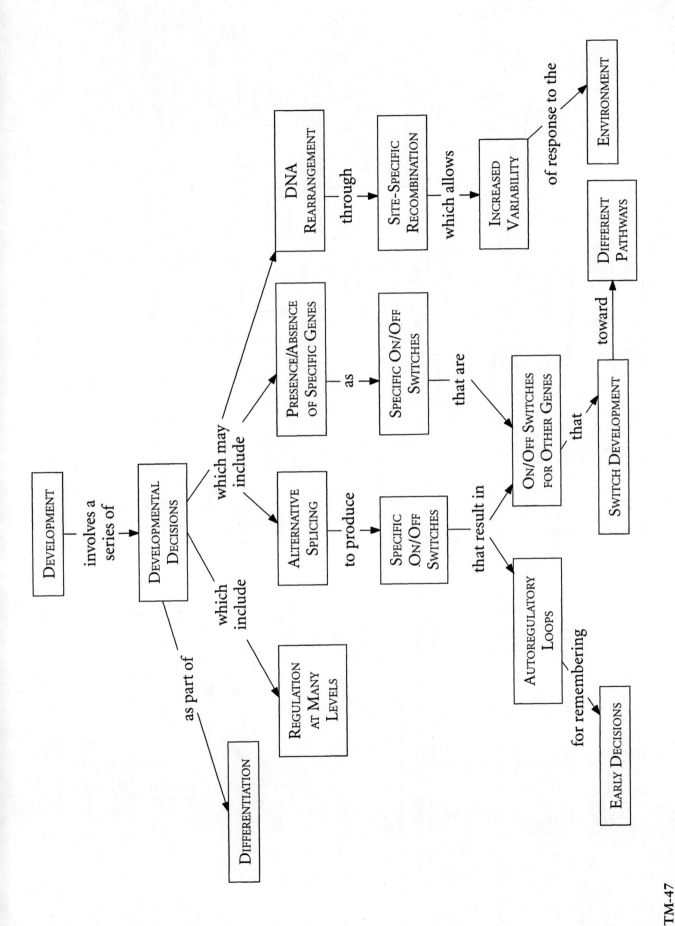

TM-47

Concept Map 23-2, page 392 in Lavett: *Student Companion with Complete*
Solutions for IGA, 6e
Copyright © 1997 W. H. Freeman and Company

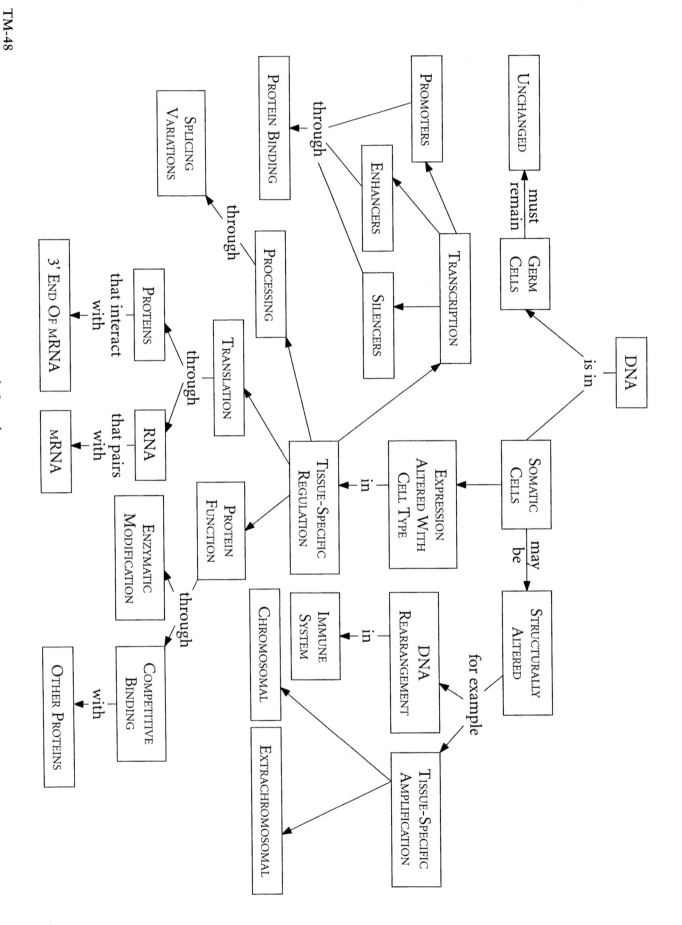

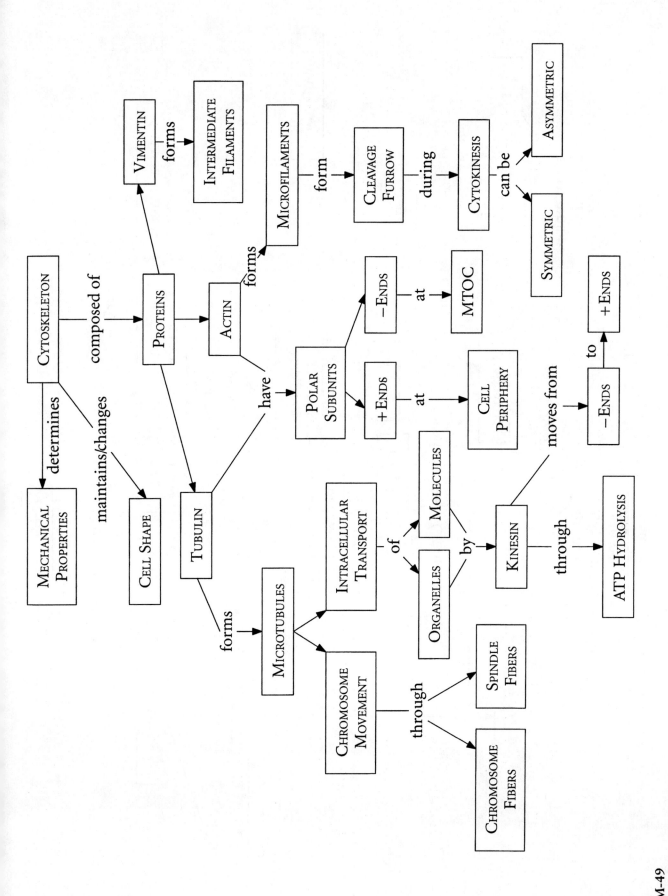

TM-50
Concept Map 24-2, page 402 in Lavett: *Student Companion with Complete Solutions for IGA, 6e*
Copyright © 1997 W. H. Freeman and Company

4 Major Divisions

Cell Cycle — has → **Positive Regulation**, **Negative Regulation**

Cell Cycle — can have → **Cell Division Cycle Mutants**

Cell Division Cycle Mutants — lead to → **Malignancy**

through → that control → **Transition**

Positive Regulation / **Negative Regulation** — of

Transition — by a → **Cascade** — of → **Small Steps**

Small Steps — involving → **Cyclin**

Cyclin → **Binding** → **Cyclin-Dependent Kinase**

Binding — that then → **Phosphorylate**

Phosphorylate → **Amino Acids**

Amino Acids — which → **Activate Transcription Factors**

Activate Transcription Factors — for next → **Small Steps**

DNA Damage — detects → **P53**

P53 — with damage activates → **P21**

P53 — at each step → **Small Steps**

P21 — that inhibits next → **Small Steps**

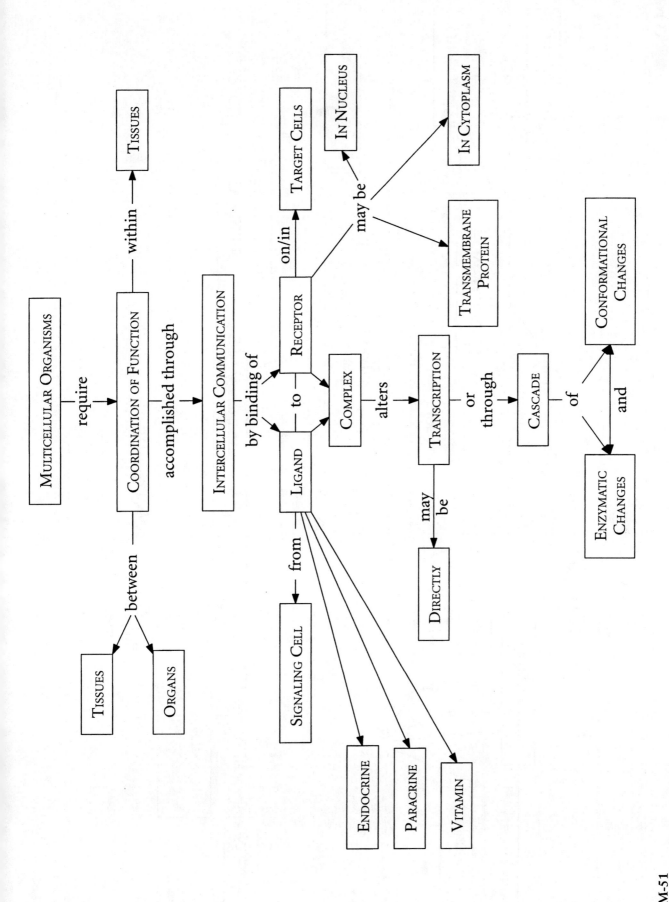

TM-51
Concept Map 24-3, page 403 in Lavett: *Student Companion with Complete Solutions for IGA, 6e*
Copyright © 1997 W. H. Freeman and Company

TM-52

Concept Map 24-4, page 405 in Lavett: *Student Companion with Complete Solutions for IGA, 6e*

Copyright © 1997 W. H. Freeman and Company

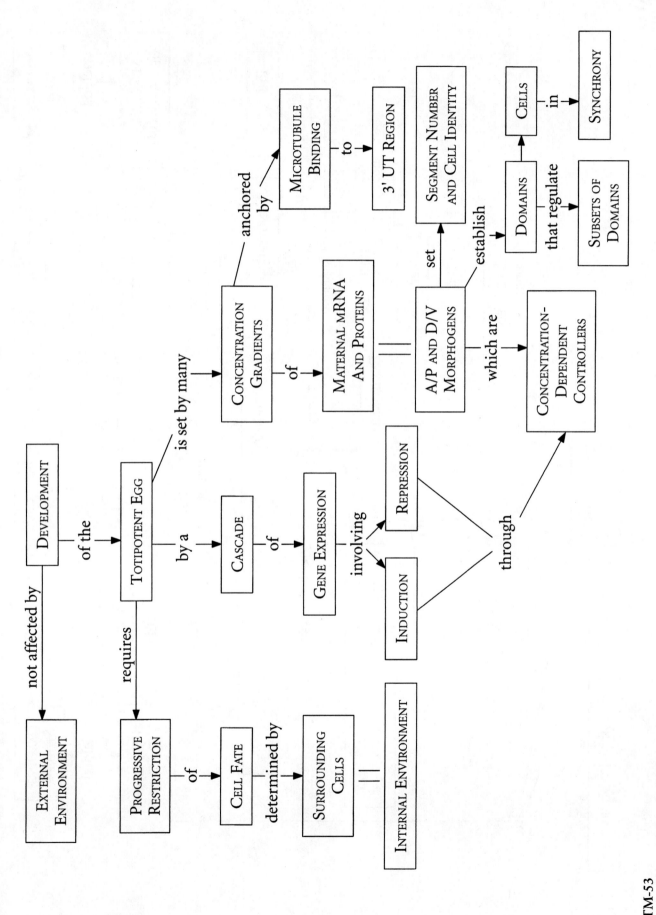

Concept Map 25-1, page 412 in Lavett: *Student Companion with Complete Solutions for IGA, 6e*
Copyright © 1997 W. H. Freeman and Company

Concept Map 26-1, page 420 in Lavett: *Student Companion with Complete*
Solutions for IGA, 6e

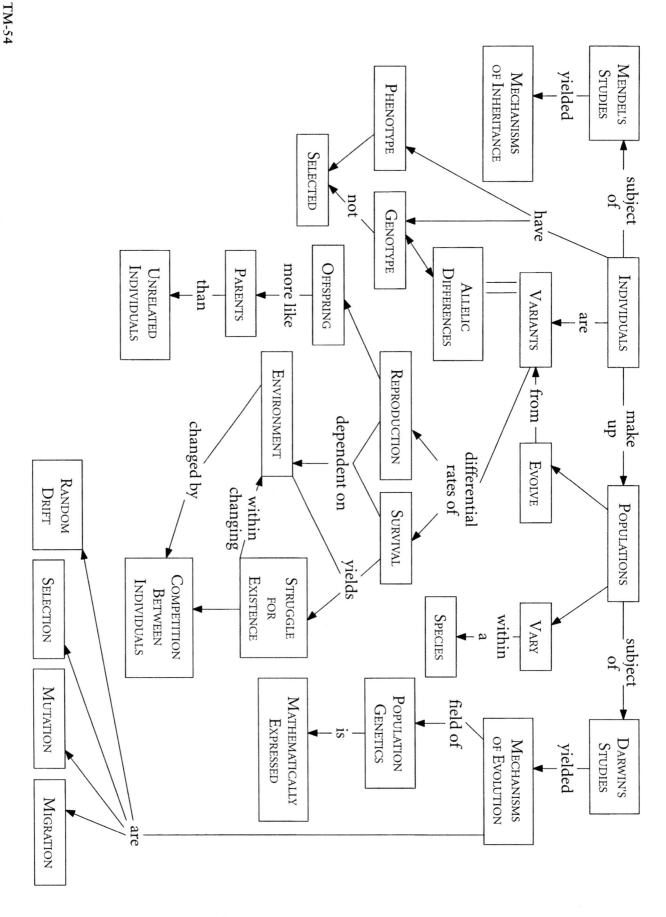

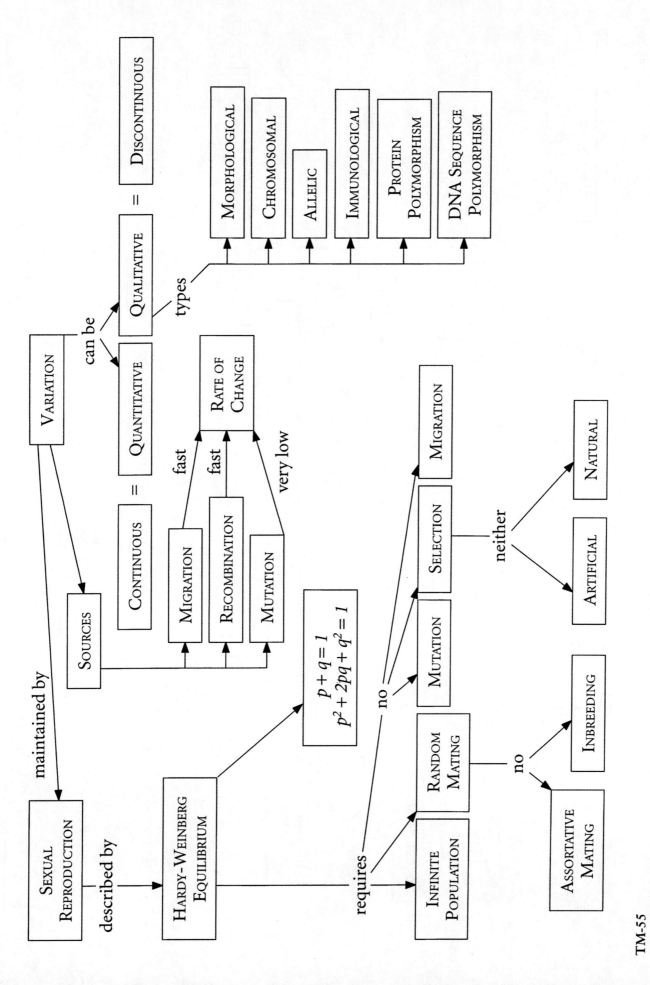

TM-55

Concept Map 26-2, page 423 in Lavett: *Student Companion with Complete*
Solutions for IGA, 6e
Copyright © 1997 W. H. Freeman and Company

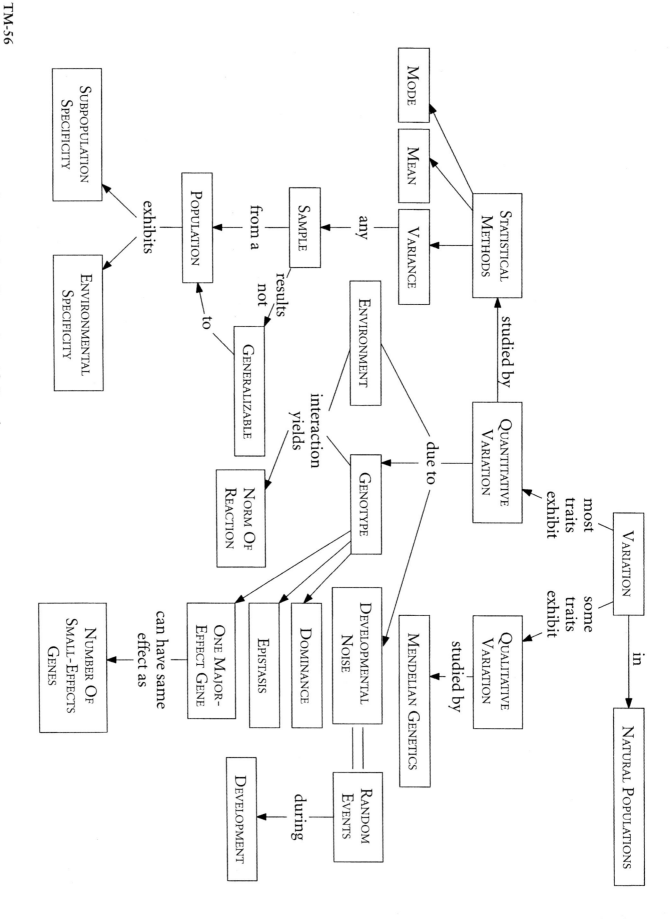